History, Developments and Trends in the Heat Treatment of Steel

History, Developments and Trends in the Heat Treatment of Steel

Editor

Peter Jurči

MDPI • Basel • Beijing • Wuhan • Barcelona • Belgrade • Manchester • Tokyo • Cluj • Tianjin

Editor
Peter Jurči
Slovak University of
Technology in Bratislava
Slovakia

Editorial Office
MDPI
St. Alban-Anlage 66
4052 Basel, Switzerland

This is a reprint of articles from the Special Issue published online in the open access journal *Materials* (ISSN 1996-1944) (available at: https://www.mdpi.com/journal/materials/special_issues/heat_treatment).

For citation purposes, cite each article independently as indicated on the article page online and as indicated below:

LastName, A.A.; LastName, B.B.; LastName, C.C. Article Title. *Journal Name* **Year**, *Volume Number*, Page Range.

ISBN 978-3-0365-0060-7 (Hbk)
ISBN 978-3-0365-0061-4 (PDF)

Contents

About the Editor

Peter Jurči, Professor, born 10 May 1968. Professor since 2010 in the field Materials Science and Engineering at the Czech Technical University in Prague. Professional orientation: heat treatment; thermo-chemical heat treatment; distortion of components due to heat- and thermo-chemical treatments; physical methods of coating tools with chromium nitride, with the addition of silver nanoparticles, cryogenic processing of tool steels, boriding of tool steels

Editorial

History, Developments and Trends in the Heat Treatment of Steel

Peter Jurči

Faculty of Materials Science and Technology in Trnava, Institute of Materials Science, Slovak University of Technology in Bratislava, 917 24 Trnava, Slovakia; p.jurci@seznam.cz

Received: 28 August 2020; Accepted: 8 September 2020; Published: 9 September 2020

Abstract: Ferrous alloys (steels and cast irons) and their heat treatment have attracted a great amount of basic and applied research due to their decisive importance in modern industrial branches such as the automotive, transport and other industries. Heat treatment is always required for these materials, in order to achieve the desired levels of strength, hardness, toughness and ductility. Over the past decades, many advanced heat- and surface-treatment techniques have been developed such as heat treatment in protective atmospheres or in vacuum, sub-zero treatment, laser/electron beam surface hardening and alloying, low-pressure carburizing and nitriding, physical vapour deposition and many others. This diversity of treatment techniques used in industrial applications has spurred a great extent of research efforts focused on the optimized and/or tailored design of processes in order to promote the best possible utilization of material properties. This special journal issue contains a collection of original research articles on not only advanced heat-treatment techniques—carburizing and sub-zero treatments—but also on the microstructure–property relationships in different ferrous alloys.

Keywords: grade 92 steel weldment; post-welding heat treatment; tensile straining; hydrogen embrittlement; ledeburitic tool steels; carburizing; rare-earth element pre-implantation; sub-zero treatments; microstructure; hardness; toughness; microstructure; fractography

---

Advanced heat- and surface-treatment techniques play a dominant role in material processing in modern industry because of the high level of protection against unwanted surface defects such as oxidation, decarburization, and too-high shape and dimensional distortion, thus reducing the final operations that are often necessary in order to correct these phenomena. Among these processes, vacuum heat treatment became of great importance in the processing of tool steels and other high-alloyed materials, owing to the fact that it produces clean surfaces after heat treatment and that the vacuum furnaces can be directly incorporated into the manufacturing lines [1]. Thermo-chemical processes such as carburizing or nitriding often use atmospheres with hazardous environmental and health impacts such as endo-gases or ammonia. Advanced thermo-chemical processing techniques such as low-pressure carburizing or nitriding, on the other hand, produce much lower amounts of environmentally dangerous substances, produce clean surfaces on tools and components, and reduce their distortion [2–4]. Moreover, these processes manifest much better efficiency, i.e., they enable the production of surface regions with greater thickness and better uniformity, with substantially shorter processing times.

In production of hard ceramic layers, processes such as chemical vapour deposition (CVD), which often produce poisonous hydrogen chloride as a by-product, were replaced by much more environmentally friendly physical vapour deposition processes. These treatment techniques, in addition, enable the design and manufacture of tailored-to-customer thin-film architecture, composition and thickness [5].

Research efforts have also been focused on the design and utilization of the high-power density treatment of metallic surfaces. Among the techniques, laser- and electron-beam surface hardening, remelting and alloying became of the greatest importance in research, development and industry. This is owing to the possibility to set up the processing parameters exactly in order to obtain the desired thickness, microstructure and related properties of treated surfaces [6].

This Special Issue contains five original, full-length articles on the effects of surface quality on the mechanical properties of hard steels, on sub-zero treatments and their impact on microstructure and mechanical properties, and on the advanced carburizing technique.

Jurči [7] reported on the worsening of the bulk toughness of hard tool steels, which results not only from increased surface roughness but also from the application of thermo-chemical treatments such as carburizing, nitriding or boronizing. On the other hand, almost no effect of physical vapour deposited (PVD) hard coatings on toughness is reported. The material toughness is exactly quantified by means of measurements of flexural strength and by the estimation of the plastic work of fracture. Toughness measurements are complemented by a thorough analysis of the fractured surfaces.

In the second paper [8], Razavykia, Delprete and Baldissera provide a comprehensive review on the cryogenic treatment of metallic materials. They discuss the improvement of material properties and durability by means of microstructural alteration comprising phase transfer, particle size adjustment, and distribution. These effects are almost permanent and irreversible. However, while improvements in the properties of materials after cryogenic treatment are discussed by the majority of reported studies, the correlation between microstructural alteration and the mechanical properties is unclear to date. At the end of the paper, the development and the trends for future research in this field are outlined and discussed.

The third paper [9] deals with the carburizing of 20Cr2Ni4A alloy steel, which was pre-implanted with either lanthanum or yttrium. The obtained results showed that rare-earth elements promoted the formation of add-on dislocations in the surface areas, which increased the carbon diffusion coefficient and thereby contributed to better carbon distribution in the carburized layers. This was reflected in the improved hardness of the low-pressure carburized components. At the end of the paper, it is stated that yttrium acts better in terms of obtaining increased surface hardness as compared to the lanthanum ion.

Kusý et al. [10] treated the Vanadis 6 steel at the cryogenic temperature of −75 °C, for different durations. They arrived at the principal findings that this kind of treatment reduces the retained austenite amount to one third as compared with that in the same steel after room-temperature quenching. Moreover, treatment at −75 °C produces a great number of extra small globular carbides in the material microstructure. This improves the prior-to-tempering hardness. However, sub-zero treatment at −75 °C leads to the complete loss of the secondary hardness peak, and the as-tempered bulk hardness manifests clear lowering. The material toughness was slightly deteriorated when the steel was low-temperature tempered, but an improvement was recorded after high-temperature tempering. The obtained results were compared with those after sub-zero treatments at −140, −196 or −269 °C.

In the fifth paper [11], the effects of the electrolytic hydrogen charging of T92 steel weldments on their room-temperature tensile properties were investigated and discussed. The weldments were differently heat treated after the welding procedure—either tempered below the transformation A_1 temperature or normalized (i.e., austenitized above the Ac_3 critical transformation temperature and subsequently air cooled) and tempered. The obtained results indicated higher hydrogen embrittlement susceptibility for the normalized-and-tempered weldments, compared to the tempered-only ones. The obtained findings were correlated with performed microstructural and fractographic observations.

All the published articles were reviewed by recognized experts in the appropriate fields through a single-blind peer-review process. As Guest Editor, I would like to acknowledge all of the authors for their valuable contributions to the Special Issue. I would also like to thank the reviewers for their comments and suggestions that greatly improved the quality of the papers. Finally, I would like to thank the Section Managing Editor, Ms. Ariel Zhou, for her kind assistance in the preparation of the Special Issue of the journal.

Funding: This research received no external funding.

Conflicts of Interest: The authors declare no conflict of interest.

References

1. Browne, R.J. A review of the fundamentals of vacuum metallurgy. *Vacuum* **1971**, *21*, 13. [CrossRef]
2. Gorockiewicz, R. The kinetics of low-pressure carburizing of alloy steels. *Vacuum* **2011**, *86*, 448. [CrossRef]
3. Jurči, P.; Stolař, P.; Šťastný, P.; Podkovičak, J.; Altena, H. Investigation of Distortion Behaviour of Machine Components due to Carburising and Quenching. *HTM J. Heat Treat. Mater.* **2008**, *63*, 27.
4. Musil, J.; Vlček, J.; Růžička, M. Recent progress in plasma nitriding. *Vacuum* **2000**, *59*, 940. [CrossRef]
5. König, U. Deposition and properties of multicomponent hard coatings. *Surf. Coat. Technol.* **1987**, *33*, 91. [CrossRef]
6. Gnanamuthu, D.S. Laser Surface Treatment. *Opt. Eng.* **1980**, *19*, 195783. [CrossRef]
7. Jurči, P. Effect of Different Surface Conditions on Toughness of Vanadis 6 Cold Work Die Steel—A Review. *Materials* **2019**, *12*, 1660. [CrossRef] [PubMed]
8. Razavykia, A.; Delprete, C.; Baldissera, P. Correlation between Microstructural Alteration, Mechanical Properties and Manufacturability after Cryogenic Treatment: A Review. *Materials* **2019**, *12*, 3302. [CrossRef]
9. Li, G.; Li, C.; Xing, Z.; Wang, H.; Huang, Y.; Guo, W.; Liu, H. Study of the Catalytic Strengthening of a Vacuum Carburized Layer on Alloy Steel by Rare Earth Pre-Implantation. *Materials* **2019**, *12*, 3420. [CrossRef]
10. Kusý, M.; Rízeková-Trnková, L.; Krajčovič, J.; Dlouhý, I.; Jurči, P. Can Sub-zero Treatment at −75 °C Bring Any Benefits to Tools Manufacturing? *Materials* **2019**, *12*, 3827. [CrossRef]
11. Čiripová, L.; Falat, L.; Homolová, V.; Džupon, M.; Džunda, R.; Dlouhý, I. The Effect of Electrolytic Hydrogenation on Mechanical Properties of T92 Steel Weldments under Different PWHT Conditions. *Materials* **2020**, *13*, 3653. [CrossRef]

Article

The Effect of Electrolytic Hydrogenation on Mechanical Properties of T92 Steel Weldments under Different PWHT Conditions

Lucia Čiripová [1], Ladislav Falat [1,*], Viera Homolová [1], Miroslav Džupon [1], Róbert Džunda [1] and Ivo Dlouhý [2]

[1] Institute of Materials Research, Slovak Academy of Sciences, Watsonova 47, 04001 Košice, Slovakia; lciripova@saske.sk (L.Č.); vhomolova@saske.sk (V.H.); mdzupon@saske.sk (M.D.); rdzunda@saske.sk (R.D.)

[2] Institute of Physics of Materials, CEITEC-IPM, Czech Academy of Sciences, Zizkova 22, 61662 Brno, Czech Republic; idlouhy@ipm.cz

*   Correspondence: lfalat@saske.sk; Tel.: +421-55-792-2447

Received: 22 July 2020; Accepted: 17 August 2020; Published: 18 August 2020

Abstract: In the present work, the effects of electrolytic hydrogen charging of T92 steel weldments on their room-temperature tensile properties were investigated. Two circumferential weldments between the T92 grade tubes were produced by gas tungsten arc welding using the matching Thermanit MTS 616 filler material. The produced weldments were individually subjected to considerably differing post-welding heat treatment (PWHT) procedures. The first-produced weldment was conventionally tempered (i.e., short-term annealed below the Ac_1 critical transformation temperature of the T92 steel), whereas the second one was subjected to its full renormalization (i.e., appropriate reaustenitization well above the T92 steel Ac_3 critical transformation temperature and subsequent air cooling), followed by its conventional subcritical tempering. From both weldments, cylindrical tensile specimens of cross-weld configuration were machined. The room-temperature tensile tests were performed for the individual welds' PWHT states in both hydrogen-free and electrolytically hydrogen-charged conditions. The results indicated higher hydrogen embrittlement susceptibility for the renormalized-and-tempered weldments, compared to the conventionally tempered ones. The obtained findings were correlated with performed microstructural and fractographic observations.

Keywords: grade 92 steel weldment; post-welding heat treatment; tensile straining; hydrogen embrittlement; metallography and fractography

---

1. Introduction

The 9 wt.% Cr creep strength enhanced ferritic (CSEF) steels (e.g., T/P91, T/P92, T/P911, C/FB2, MARBN, NPM1, etc.) represent advanced structural materials for application in high-efficiency power engineering. However, for constructing complex power generation equipment, fusion welding technologies are needed for joining individual functional parts. In accordance with the numerous research studies and ex-service experience, e.g., [1–5], it has been generally accepted that the fusion welded joints of ferritic steels represent the most critical component locations with respect to their preferential degradation and potential failure. Besides the regions of base material (BM) and weld metal (WM) within the structures of all welded joints, thermal effect of fusion welding on the welded ferritic steels' BMs typically results in the creation of a relatively wide heat-affected zone (HAZ) consisting of several, continuously created microstructural sub-regions, i.e., often called the "HAZ microstructural gradient". Its occurrence within the welded joint represents the primary, welding-induced microstructure degradation zone, since the individual HAZ sub-regions, such as the coarse-grained HAZ (CG-HAZ), fine-grained HAZ (FG-HAZ), inter-critical HAZ (IC-HAZ), and subcritical HAZ (SC-HAZ), possess mutually various microstructures and mechanical properties [6–8].

Depending on several factors including the welding metallurgy-related material properties and outer loading and/or environmental conditions, the fusion weldments can be susceptible to some of several typical failures [9]. The "Type I" and "Type II" failures, originating from intercrystalline cracks in weld metals, are generally related to the so-called "hot cracking" phenomena, typically occurring in weld solidified microstructures with higher impurity content. However, the occurrence of these failures has been considerably suppressed in the ferritic steels' weldments thanks to the recently developed ferritic filler materials of high metallurgical purity [10]. The "Type III" failure typically occurs within the CG-HAZ close to the weld fusion zone (FZ) of low alloy ferritic steel weldments. This failure type has been often related to the so-called "reheat cracking" due to either residual stress relief during the PWHT or superabundant secondary precipitation hardening in FZ/CG-HAZ during high temperature creep exposure [11–13]. In a specific case of dissimilar weldments, the considered failure type (sometimes referred to as "Type IIIa" failure [14]) is related to premature creep cracking within the area of soft, carbon-depleted CG-HAZ, created as a result of the decarburization processes driven by the carbon activity gradient at the interface between the lower grade ferritic steel and the higher grade weld metal. Last but not least, depending on acting environmental conditions, the "Type III" failure may also be related to so-called "cold cracking" phenomena, i.e., hydrogen-induced cracking (HIC) or environmentally assisted cracking (EAC) [15,16]. This failure occurs due to exceeding the critical hydrogen concentration in locally hardened FZ/CG-HAZ areas with the highest degree of transformation (i.e., martensitic) hardening as a consequence of the welding thermal cycle. Under long-term creep conditions, the welded joints of ferritic heat-resistant steels are typically prone to the "Type IV" failure within their FG-/IC-HAZs because these regions exhibit the lowest creep strength within the whole weldment. This failure is generally related to severe degradation of transformation hardening mechanism and preferential coarsening of $Fe_2(W,Mo)$-based Laves phase within the failure location [17,18]. The study by Albert et al. [19] showed that the "Type IV" failure is caused by preferential creep strain accumulation in the soft, fine-grained HAZ regions (FG-/IC-HAZs) due to the multiaxial stress state induced by microstructural heterogeneity throughout the weld-joint. A specific failure type is related to "cracking in over-tempered base material" which typically occurs within the softened region of SC-HAZ and is characterized by a highly ductile fracture [20,21]. This failure type is observed usually in welded joints after the high temperature tensile tests or after high-stress short-term creep tests [21,22]. The mechanism of microstructural and property degradation in SC-HAZ, i.e., within the over-tempered base material, is believed to arise from the coarsening of precipitates during the welding thermal cycle. After longer durations of low-stress creep tests, the failure commonly shifts from the over-tempered region to the "Type IV" failure region. However, unlike the short-term "over-tempered base metal cracking", the long-term "type IV cracking" is characterized by low-ductility creep failure [23].

In common industrial practice, the weldments of CSEF steels are necessarily subjected to conventional PWHT, i.e., the subcritical tempering below the steel Ac1 critical transformation temperature. The main aim of such PWHT is to relieve residual stresses and thermally stabilize the weld microstructure with secondary phase precipitates, typically the $M_{23}C_6$ (M = Cr, Fe …) carbides and MX (M = V, Nb; X = C, N) carbo-nitrides. The direct consequence of performing the subcritical PWHT procedure is related to the decrease of unallowably high hardness in FZ/CG-HAZ and improvement of the overall weld fracture resistance [19,24]. However, it has been proved [19] that the occurrence of premature "Type IV" creep failure in ferritic steels' weldments cannot be avoided by any variation in the subcritical PWHT regime, since their HAZ microstructural gradients remain still preserved within subcritically tempered microstructures. On the other hand, several studies [25,26] suggested that the only way to enable the "Type IV" failure suppression in ferritic welds is associated with so-called "full heat treatment" which involves the weld renormalization (i.e., the weld complete reaustenitization and its subsequent cooling on still air), followed by conventional subcritical tempering.

Our previous investigations [27,28] were focused on investigation of the effects of both the conventional tempering and quenching-and-tempering PWHT procedures of T92/TP316H martensitic/austenitic weldments on their microstructure and creep behavior. The results showed that the quenching-and-tempering PWHT led to "Type IV" failure elimination and thus notable creep life improvement as a result of significant homogenization of the T92 steel microstructure, i.e., complete suppression of the T92 HAZ microstructural gradient thanks to performed reaustenitization. Moreover, our separate study [29] on the T92 HAZ local mechanical properties of the T92/TP316H weldments indicated, that compared to the weldments subjected to only conventional PWHT, the T92 HAZ of quenched-and-tempered weldments exhibited lower hardness and higher impact toughness. The combined effects of quenching-and-tempering PWHT and subsequent electrochemical hydrogen charging on room-temperature tensile properties of the T92/TP316H weldments were investigated in [30]. It has been revealed that the applied electrochemical hydrogen charging did not affect the strength properties of the weldments significantly, but it resulted in quite serious deterioration of their deformation properties along with significant impact on their fracture behavior and final failure localization. The most critical region was found to be the interfacial weld region close to the T92 steel FZ.

Our present study represents a continuous research work to our aforementioned former studies. It deals with investigation of the effects of initial PWHT conditions and subsequent electrochemical hydrogenation on the resulting room-temperature tensile properties and fracture behavior of T92/T92 welded joints. Mutual correlations between varying microstructural characteristics induced by different initial PWHT regimes and resulting mechanical properties of the weldments in either hydrogen-free or hydrogen-charged conditions are discussed.

2. Materials and Methods

Four segments of industrially normalized and tempered T92 tubes (outer diameter 38 mm, wall thickness 5.6 mm, approx. tube segment length 130 mm) were circumferentially welded in the company SES a.s. Tlmače, Slovakia. The welded joints were produced by gas tungsten arc welding (GTAW) technique using T92-based filler metal Thermanit MTS 616 to prepare two equivalent T92/T92 weldments. The T92/T92 welds geometry was the same as also used in our previous study about long-term ageing effects on room-temperature tensile behavior of quenched and tempered T92/TP316H dissimilar weldments [31]. Specifically, the 60° groove angle and 2–3 mm root gap was used. Welding parameters for the preparation of T92/T92 welded joints were the following ones: welding current 120–160 A, voltage 12–17 V and heat input 9–12 kJ/cm. The diameter of TIG electrode was 2.4 mm and the negative polarity on the electrode was used. Table 1 shows chemical compositions of the T92 steel base material (T92 BM) and T92 steel-based filler metal (T92 FM) Thermanit MTS 616.

Table 1. Chemical composition (wt.%) of T92 base material (T92 BM) and T92-based filler metal (T92 FM) used for fabrication of T92/T92 weldments.

Material	C	N	Si	Mn	Cr	Mo	W	B	Ni	Ti	V	Nb	Fe
T92 BM	0.11	0.05	0.38	0.49	9.08	0.31	1.57	0.002	0.33	-	0.2	0.07	rest
T92 FM	0.11	0.05	0.2	0.6	8.8	0.5	1.6	-	0.7	-	0.2	0.05	rest

The chemical compositions in Table 1 represent certified alloy compositions by the material producers Tenaris Dalmine (Dalmine—BG, Italy) and Voestalpine Böhler Welding (Düsseldorf, Germany), respectively.

The two prepared weldments were individually subjected to mutually differing post-welding heat treatment (PWHT) procedures. Figure 1 shows schematic illustration of both these PWHT procedures in context with the equilibrium phase diagram including isoplethal section for T92 BM, computed by thermodynamic software ThermoCalc (version S, Thermo-Calc Software AB, Solna, Sweden) using thermodynamic database TCFE6.

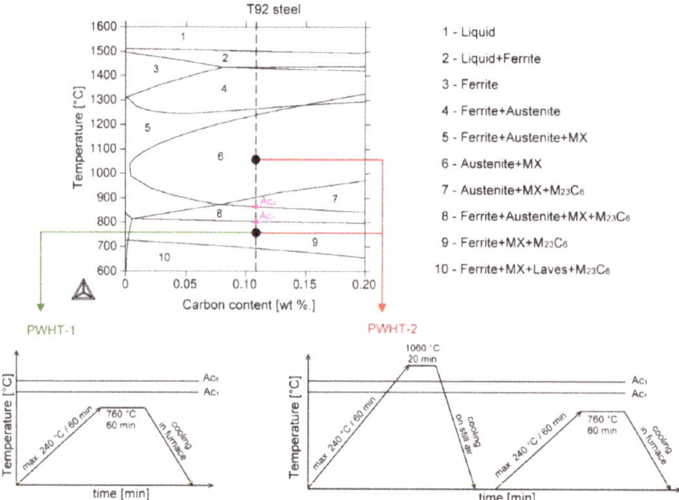

Figure 1. Calculated equilibrium phase diagram with schematic illustrations of individual post-welding heat treatment (PWHT) regimes applied in the present study for T92/T92 weldments (The T92 steel composition is indicated in the diagram by vertical dashed line at 0.11 wt.% C).

The first T92/T92 weldment was conventionally tempered at 760 °C (i.e., below the Ac_1 temperature of T92 steel) for 60 min and then slowly cooled within the tempering furnace (see the PWHT-1 in Figure 1). On the other hand, the second T92/T92 weldment was subjected to its full renormalization consisting of the complete reaustenitization at 1060 °C (i.e., well above the Ac_3 temperature of T92 steel) for 20 min and subsequently cooled on still air, followed by its conventional subcritical tempering (see the PWHT-2 in Figure 1). From both weldments, twelve cylindrical tensile test specimens of cross-weld (c-w) configuration with partly discontinuous M6 thread (due to the above specified tube wall thickness) within their head portions were machined. A schematic illustration of the tensile test specimen is shown in Figure 2.

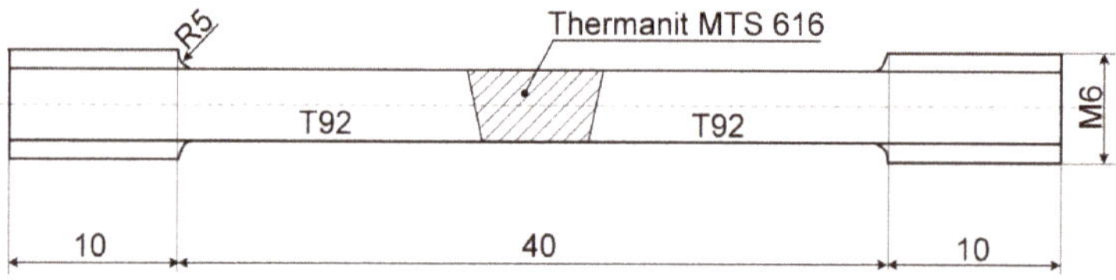

Figure 2. The tensile test specimen for cross-weld tensile testing of T92/T92 weldments (All dimensions are in mm), gauge length diameter 4 mm, gauge length 38 mm.

Electrolytic hydrogenation, i.e., cathodic hydrogen charging of prepared cylindrical c-w tensile specimens was performed in electrolytic solution of 1M HCl with 0.1N $N_2H_6SO_4$ at a current density of 300 A/m². The hydrogenation was realized at room temperature for 24 h. This procedure has been optimized and used in our several former studies [15,30,32] which indicated full saturation of tensile specimens by hydrogen after 24 h of their electrolytic hydrogenation. Similar findings, supported by hydrogen concentration measurements indicated the same or even shorter hydrogenation time for achieving the hydrogen concentration saturation in electrochemically hydrogen-charged alloy steels, as reported in other studies, e.g., [33–35]. Yin et al. [36] indicated that the content of diffusible hydrogen tends to be the saturation state when the hydrogen charging time reaches 48 h. However, they showed that the difference in diffusible hydrogen concentration for 24 and 48 h of hydrogen charging was already rather small (i.e., within experimental value scattering). A schematic illustration of the whole experimental setup is visualized in Figure 3.

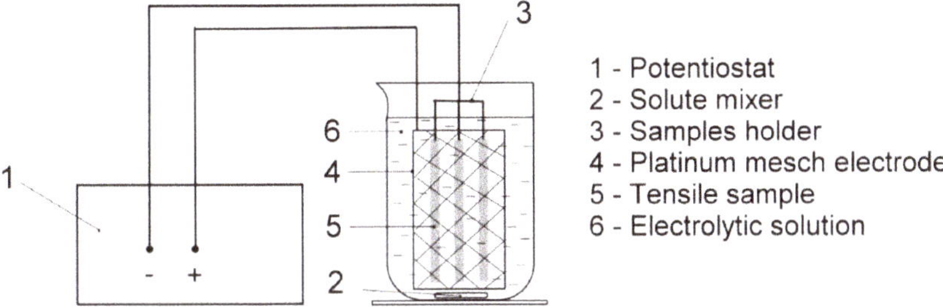

Figure 3. Schematic illustration of electrolytic hydrogenation.

The room-temperature tensile tests were performed for individual welds' PWHT states in both hydrogen-free and hydrogen-charged conditions. The tensile testing was carried out using TIRATEST 2300 universal testing machine (TIRA GmbH, Schalkau, Germany) at a crosshead speed of 0.05 mm/min. Three tensile test specimens per each state (i.e., "PWHT-1", "PWHT-2", "PWHT-1 + hydrogen", and "PWHT-2 + hydrogen") were investigated. The hydrogen-charged samples were tested immediately after the electrolytic hydrogen charging. The evaluation of c-w tensile properties (i.e., yield stress "YS" estimated as 0.2% proof stress, ultimate tensile strength "UTS", total elongation at fracture "EL", and reduction of area at fracture "RA") involved the calculation of their average values and corresponding standard deviations.

Local mechanical properties of studied weldments were characterized by means of hardness measurements which were performed using a Vickers 432 SVD hardness tester (Wolpert Wilson Instruments, division of Instron Deutschland GmbH, Aachen, Germany) on plain surfaces of longitudinal sections of fractured tensile specimens. This procedure was also helpful for indication of local strain hardening effects within the studied weldments during the tensile tests. The referential, i.e., un-deformed samples corresponding to both initial PWHT states were also tested for hardness. All the hardness measurements were performed at 98 N loading for 10 s per measurement.

Microstructural analyses of the studied weldments were performed on the conventionally prepared metallographic samples (i.e., wet grinding on SiC papers with granularity from 500 to 1200 grit, cloth polishing with a diamond paste suspension of a particle size ranging from 1 to 0.25 μm and final etching in a solution consisting of 120 mL CH_3COOH, 20 mL HCl, 3 g picric acid, and 144 mL CH_3OH) using the light optical microscope OLYMPUS GX71 (OLYMPUS Europa Holding GmbH, Hamburg, Germany) and the scanning electron microscope (SEM) JEOL JSM-7000F (Jeol Ltd., Tokyo, Japan). Fractographic analyses were carried out using the SEM Tescan Vega-3 LMU (TESCAN Brno, s.r.o., Czech Republic).

3. Results and Discussion

3.1. Microstructures

Since the qualitative microstructural characteristics of T92/T92 weldments are, in principle, symmetrically distributed with respect to the weld centerline, only one half part of the cross-weld microstructure was documented. Figure 4 shows the light-optical micrograph of T92/T92 weldment in conventionally tempered, i.e., PWHT-1 material state. It can be clearly seen that the weldment after the PWHT-1 exhibits a typical microstructural gradient consisting of individual microstructural sub-regions, i.e., BM, SC-HAZ, IC-HAZ, FG-HAZ, CG-HAZ, FZ, and WM. These microstructural sub-regions are generally formed of tempered martensitic-ferritic structures with different tempering grades of martensite, depending on the reached local peak temperatures (i.e., the temperature gradient) during the welding thermal cycle.

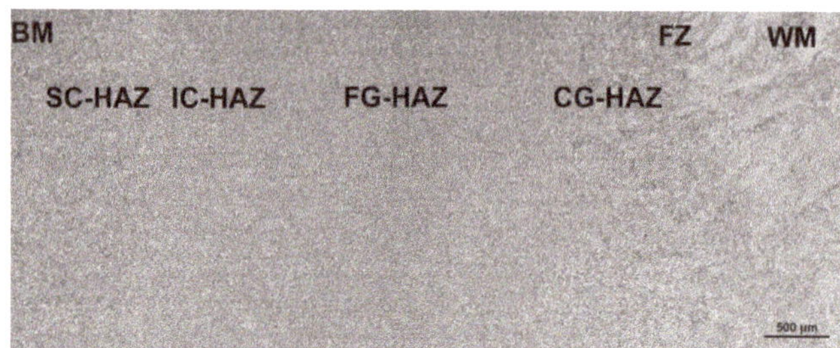

Figure 4. Light-optical micrograph of T92/T92 weldment after the tempering PWHT-1.

The microstructural transitions SC-HAZ/IC-HAZ, FG-HAZ/CG-HAZ, and FZ/WM are clear thanks to the observed differences in grain size and morphology. However, the microstructural transitions BM/SC-HAZ and IC-HAZ/FG-HAZ cannot be clearly differentiated by means of light optical microscopy and, thus, they are only roughly estimated in Figure 4. Figure 5 shows the light-optical micrograph of T92/T92 weldment in renormalized-and-tempered, i.e., PWHT-2 material state.

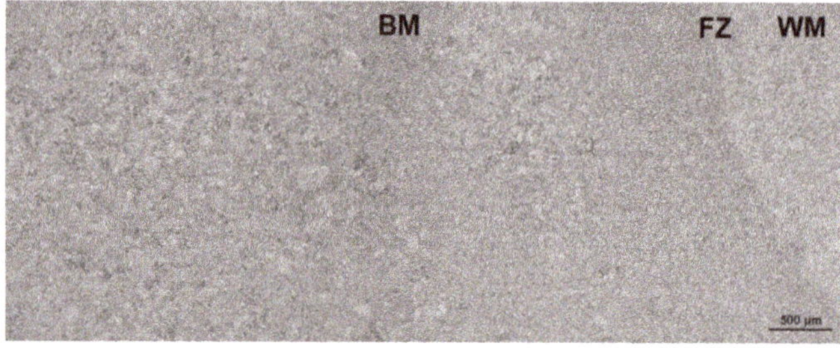

Figure 5. Light-optical micrograph of T92/T92 weldment after the renormalizing-and-tempering PWHT-2.

It can be seen that the weldment after the PWHT-2 shows quite homogenized microstructure as a consequence of the performed renormalization treatment. Within the renormalized-and-tempered weldment, only the regions of BM and WM can be clearly distinguished (Figure 5). This observation can be directly related to the homogenization effect of the applied PWHT-2 resulting in notable suppression of the original T92 HAZ microstructural gradient due to the performed renormalization. However, it should be noted that Figure 5 indicates also a certain microstructural refinement within the former CG-HAZ and WM regions compared to the microstructure of rest BM involving the renormalized-and-tempered regions of former SC-HAZ, IC-HAZ, and FG-HAZ. The detailed SEM-micrographs of individual microstructural zones of T92/T92 weldment after the PWHT-1 and PWHT-2 are shown in Figures 6 and 7, respectively.

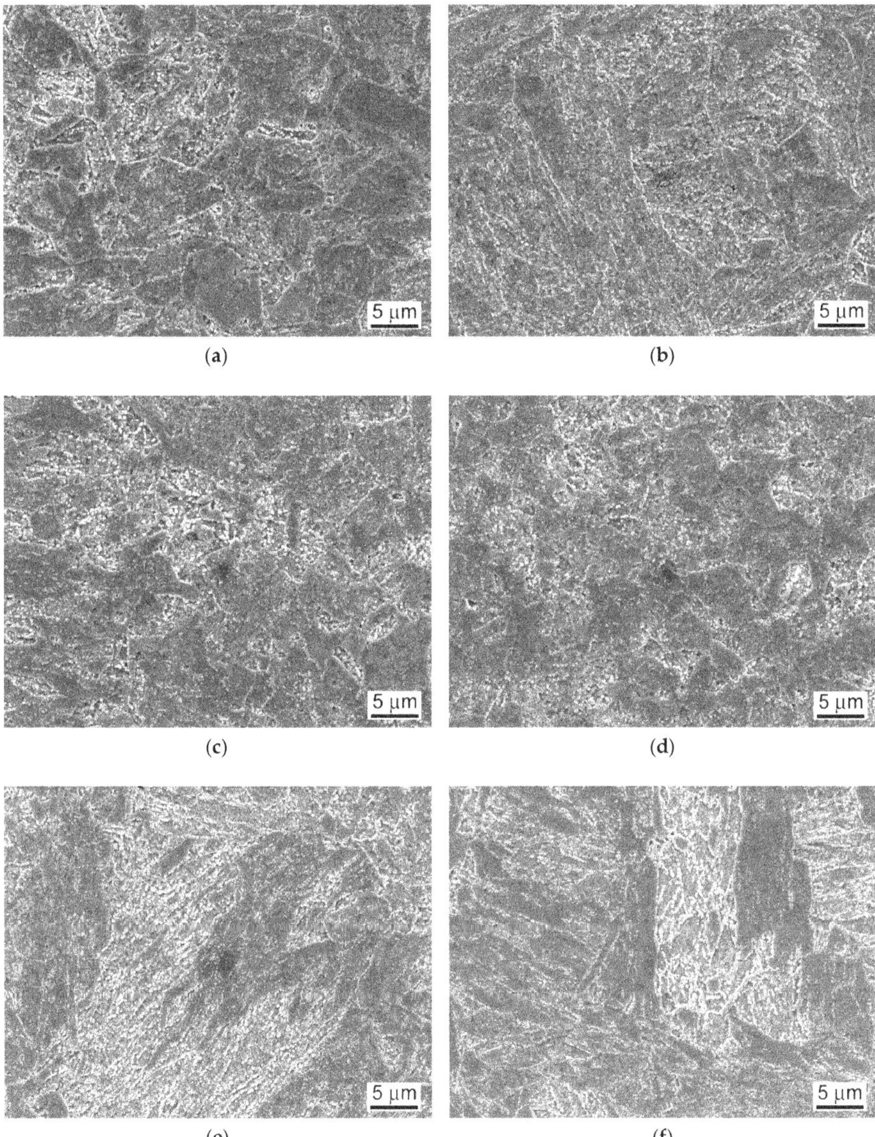

Figure 6. SEM-micrographs of individual microstructural zones of T92/T92 weldment after the tempering PWHT-1: (**a**) BM; (**b**) SC-HAZ; (**c**) IC-HAZ; (**d**) FG-HAZ; (**e**) CG-HAZ; and (**f**) WM.

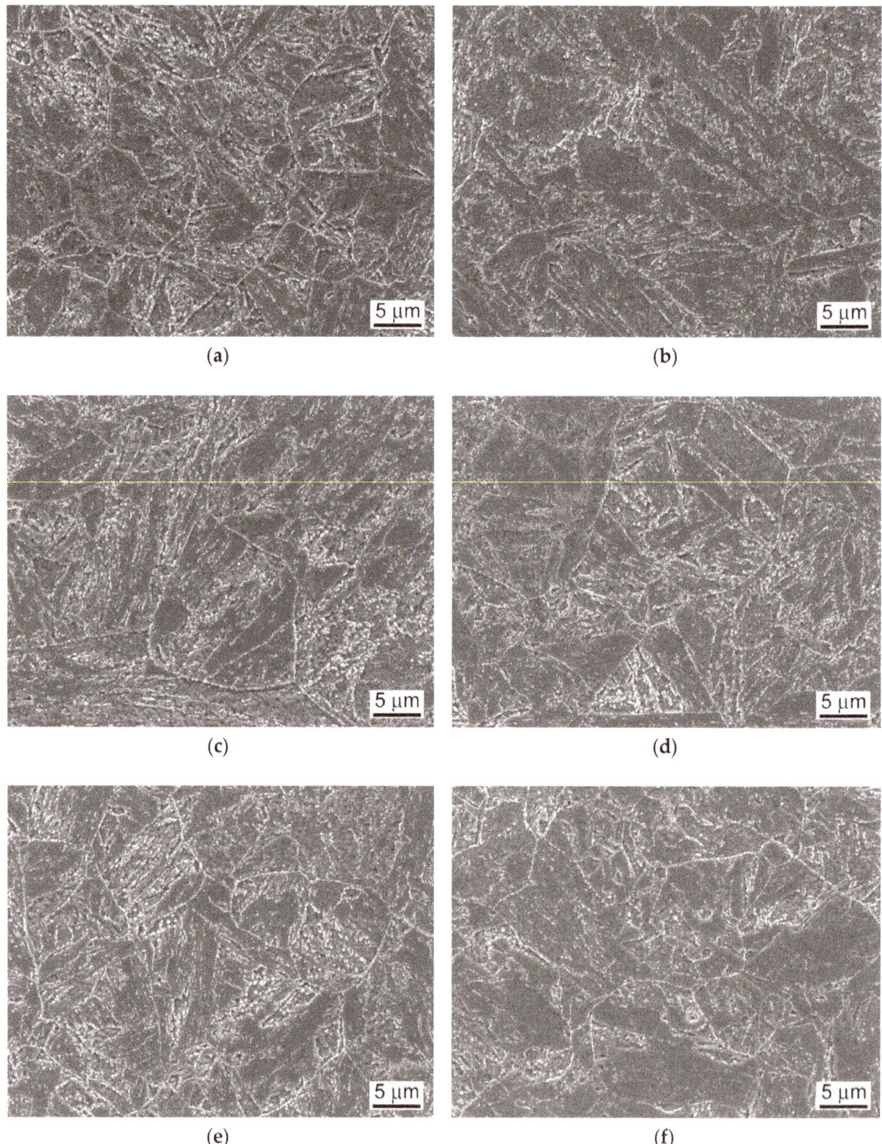

Figure 7. SEM-micrographs of individual microstructural zones of T92/T92 weldment after the renormalizing-and-tempering PWHT-2: (**a**) BM; (**b**) former SC-HAZ; (**c**) former IC-HAZ; (**d**) former FG-HAZ; (**e**) former CG-HAZ; and (**f**) WM.

The phase composition of normalized and tempered martensitic steels of T/P92 grade is generally known and consists of ferritic matrix and strengthening precipitates of intergranular $M_{23}C_6$ (M = Cr, Fe …) carbides and intragranular MX (M = V, Nb; X = C, N) carbo-nitrides [37–41]. The same phase composition is to be expected also in the currently studied T92/T92 weldments in both the conventionally tempered and renormalized-and-tempered PWHT conditions. Although predicted by the phase diagram in Figure 1, the precipitation of intermetallic $Fe_2(W,Mo)$ Laves phase is not to be expected in the currently studied material states (PWHT-1 and PWHT-2) due to insufficient time for its creation related to slow diffusion kinetics of the tungsten and molybdenum atoms in the ferrite solid solution. This assumption has already been evidenced in our several former studies [27–29] about the effect of PWHT conditions on microstructure and various properties of dissimilar T92/TP316H ferritic/austenitic weldments for high temperature applications.

By comparison of individual microstructures in Figures 6 and 7, it can be stated that the performed PWHT-2 did not induce full microstructural homogenization of studied weldment with respect to the grain size. Thus, from the observed microstructural characteristics in Figure 7, it cannot be explicitly judged about the appropriateness of the used PWHT-2. Although the original HAZ microstructural gradient has been considerably suppressed, some recognizable microstructural heterogeneity among former HAZ sub-regions is still to be observed. The originally fine-grained regions (IC-HAZ, FG-HAZ) related to PWHT-1 became notably coarse-grained after the PWHT-2. On the contrary, the originally coarse-grained regions (CG-HAZ and WM) became partly refined. The observed microstructural changes can be related to variant (non-uniform) microstructural evolution in individual sub-regions during the PWHT-2 due to pre-existing microstructural differences originated from the primary welding-induced microstructural changes. The IC-HAZ microstructure is formed in the region of BM heated up during the welding to inter-critical peak temperatures (i.e., the temperatures in Ac_1-Ac_3 range). Accordingly, the IC-HAZ is formed of fine-grained microstructure consisting of over-tempered (i.e., non-transformed) martensite (i.e., ferrite with coarsened precipitates of original undissolved carbides) and newly formed martensite created on cooling from fine-grained non-saturated austenite. The FG-HAZ microstructure is formed in the region of BM heated up during the welding to peak temperatures just above Ac_3 up to about 1100 °C. After subsequent cooling, the resulting FG-HAZ consists of newly formed martensite created from fine-grained non-saturated austenite and coarsened precipitates of original undissolved carbides. As shown in Figure 7, after the PWHT-2 both the IC-HAZ and FG-HAZ microstructures exhibit pronounced grain growth which can be related to the lower pinning effect of the coarsened carbide precipitates on the grain boundaries. Thus, the evolution of pronounced grain coarsening in both originally fine-grained IC-HAZ and FG-HAZ microstructures during the renormalization seems to be associated with their low thermal stability in as-welded material condition, enhancing the thermodynamic driving force for the observed microstructural changes. In contrast to the IC-HAZ and FG-HAZ regions, the original CG-HAZ and WM regions are created as a result of on-cooling phase transformations from the highest peak temperatures reached during the welding thermal cycle. Thanks to carbide dissolution at considered peak temperatures, these regions consist of coarse grain structures with the highest level of transformation (i.e., martensitic) hardening. During renormalization at 1060 °C (Figure 1) of these primarily coarse-grained microstructures, thermodynamic conditions for further grain growth are rather unfavorable due to a relatively low renormalization temperature compared to the considered highest peak temperatures. Instead, thanks to the effect of carbon supersaturation in these martensitic microstructures, the creation of newly formed (small) austenite grains on heating to the renormalization temperature is thermodynamically favored. This transformation behavior during the weld renormalization followed by on-cooling phase transformations and final tempering is assumed to be the reason for the creation of partly refined microstructures of CG-HAZ and WM after the PWHT-2.

3.2. Mechanical Properties

The effects of initial PWHT conditions in combination with subsequent electrolytic hydrogenation of the studied T92/T92 weldments on their room-temperature tensile properties are shown in Figure 8.

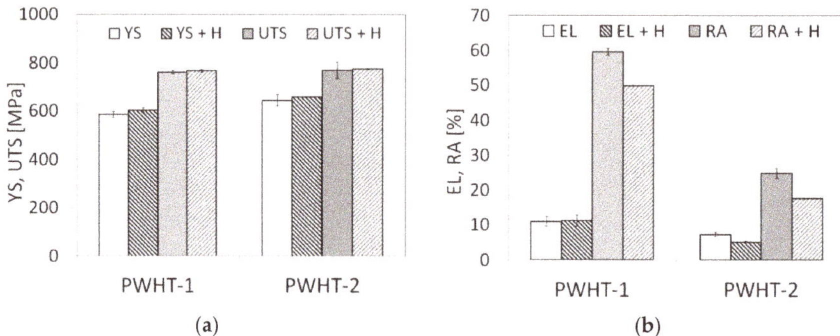

Figure 8. The effects of PWHT conditions and subsequent electrolytic hydrogenation on room-temperature tensile properties of investigated T92/T92 weldments: (**a**) strength properties and (**b**) deformation properties.

It can be seen that the renormalizing-and-tempering PWHT-2 of studied T92/T92 weldment resulted only in a small increase of the strength properties, i.e., the YS and UTS values, compared to those of the weldment after the tempering PWHT-1 (Figure 8a). Thus, it can be stated that the effect of various PWHT conditions on the resulting strength properties of studied weldment was rather insignificant. In addition, the differences in measured strength properties between the hydrogen-free and hydrogen-charged weldments in the both PWHT conditions were also quite negligible (Figure 8a). On the contrary, the renormalizing-and-tempering PWHT-2 of studied weldment resulted in a significant decrease of the deformation properties, i.e., the EL and RA values, compared to those of the weldment after the tempering PWHT-1 (Figure 8b). Additional hydrogen charging of the studied weldments in both PWHT material states led to further deterioration of their deformation properties (Figure 8b). However, the observed detrimental effect of the renormalizing-and-tempering PWHT-2 on the deformation properties of the weldments was much more pronounced in comparison with the effect of electrolytic hydrogenation. The measure of individual studied effects (i.e., the heat treatment procedure and electrolytic hydrogenation) on the plasticity deterioration can be quantitatively estimated using the so-called embrittlement index EI:

$$\mathrm{EI}\,(0,\,x)\;=\;\frac{RA_0 - RA_x}{RA_0}\times 100\%$$ (1)

where RA_0 and RA_x are the values of reduction of area at fracture of two considered material states, and the subscripts "0" and "x" refer to the states selected as initial and final, respectively [30]. Thus, the calculated values of the embrittlement index using the average RA values (Figure 8b) are summarized in Table 2. From the calculated values of embrittlement index in Table 2, it is clear that the highest degree of embrittlement of studied weldments is caused by the application of renormalizing-and-tempering PWHT-2 (row 1). When comparing the effects of additional hydrogen charging, the higher measure of hydrogen embrittlement is indicated for the weldments processed by the PWHT-2 (row 3) compared to the weldments processed by the PWHT-1 (row 2).

Table 2. Embrittlement index for individual material states.

Row	0	x	EI $(0, x)$ [%]
1	PWHT-1	PWHT-2	58.2
2	PWHT-1	PWHT-1 + H	16.1
3	PWHT-2	PWHT-2 + H	28.9

The obtained findings about the highly detrimental effect of the welds' homogenization treatment on their plastic properties can be explained by considering the microstructural changes induced by the renormalizing-and-tempering PWHT-2. The observed microstructural differences between the weldments in PWHT-1 and PWHT-2 material states (Figures 4–7) have crucial effects on the localization of plastic deformation during the tensile testing. In order to indicate local strain hardening behavior in the studied T92/T92 weldments during the tensile straining, c-w Vickers hardness measurements were carried out on the plain surfaces of longitudinal sections of fractured tensile samples after the tensile tests. Figures 9 and 10 show the c-w Vickers hardness profiles of the studied weldments initially heat treated using the PWHT-1 and PWHT-2, respectively. Visible interruptions within the c-w hardness profiles indicate the fracture locations of broken tensile test specimens after the room-temperature tensile tests. The c-w hardness profiles of the referential unstrained samples are also included for comparison. Significant differences between the c-w hardness profiles of T92/T92 weldments in various initial PWHT conditions are clearly visible when comparing Figures 9 and 10. By comparison of both unstrained samples (i.e., in PWHT-1 and PWHT-2 conditions), it can be concluded that the weldment after the tempering PWHT-1 shows a steep hardness gradient in its HAZ. The highest hardness values are measured in the WM and FZ regions which can be related to the highest measure of transformation (i.e., martensitic) hardening in these locations. On the other hand, the lowest hardness values are typically measured in the FG-HAZ, IC-HAZ, and partly SC-HAZ regions which are known to be locations with the greatest degradation of transformation hardening during the welding thermal cycle [10].

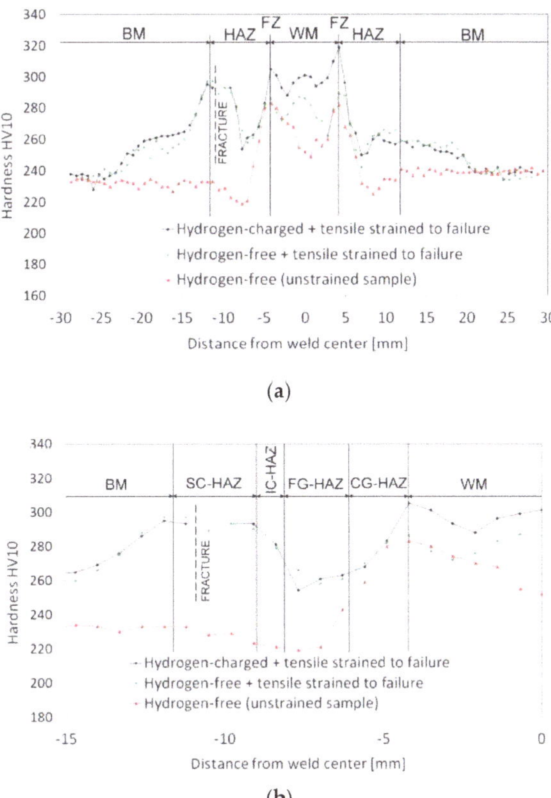

(**a**)

(**b**)

Figure 9. The effects of electrolytic hydrogenation and subsequent room-temperature tensile testing on cross-weld hardness profiles of T92/T92 weldments initially heat treated by conventional tempering PWHT-1: (**a**) overall hardness profiles and (**b**) detailed hardness profiles focused on the HAZ.

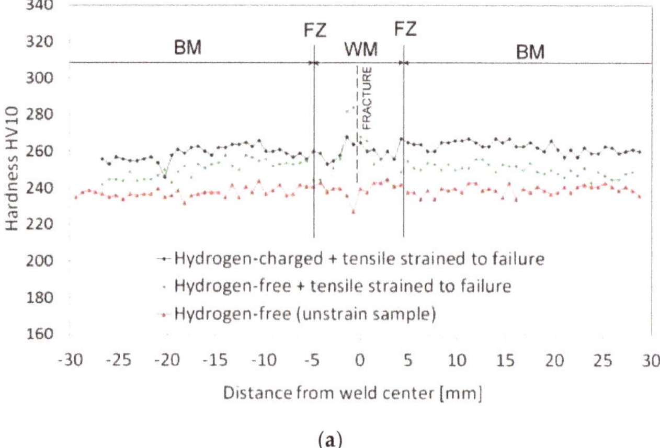

(**a**)

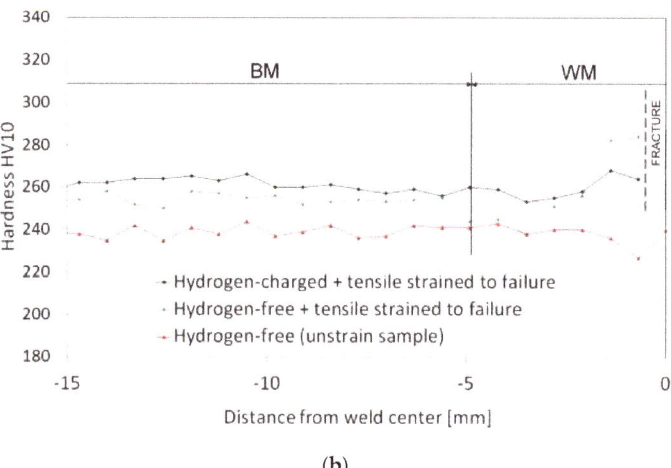

(**b**)

Figure 10. The effects of electrolytic hydrogenation and subsequent room-temperature tensile testing on cross-weld hardness profiles of T92/T92 weldments initially heat treated by renormalizing-and-tempering PWHT-2: (**a**) overall hardness profiles and (**b**) detailed hardness profiles focused on the former HAZ.

The hardness profiles of both the hydrogen-free and hydrogen-charged specimens show the strain-induced hardness peaks in the formerly soft (i.e., in the initial unstrained state) SC-HAZ in the location of tensile test fracture. Thus, the originally softer areas within the T92/T92 weldment in the PWHT-1 material state represent the locations of the localization of the highest plastic deformation during tensile testing.

The necking-related hardness peaks at fracture locations are the result of strain hardening due to the localization of plastic deformation after reaching plastic instability during tensile testing. This can be supported by the fact that the more ductile samples after the PWHT-1 (in both hydrogen-free and hydrogen-charged conditions, see Figure 9) show significant hardness peaks, compared to the less pronounced hardness peak related to the embrittled sample after the PWHT-2. Moreover, the hydrogenated sample after the PWHT-2 shows almost total suppression of its hardness peak in fracture location (see Figure 10). The localized plastic deformation can reasonably explain the better deformation properties of the highly heterogeneous T92/T92 weldment in the PWHT-1 material state, compared to the homogenized weldment in the PWHT-2 material state. The localized plastic deformation can be quantified by true fracture strain ε_f, as it follows:

$$\varepsilon_f = ln\frac{A_0}{A_f} \tag{2}$$

where A_0 and A_f are the values of original cross-sectional area and final (i.e., minimal) cross-sectional area at the point of fracture of the tensile test specimen. The calculated ε_f values indicating the level of accumulated plastic deformation at the location of final fracture are shown in Table 3:

Table 3. Calculated values of true fracture strain for studied weldment in individual material states.

Material State	ε_f (–)
PWHT-1	0.907
PWHT-1 + hydrogen	0.693
PWHT-2	0.286
PWHT-2 + hydrogen	0.195

In our case, the reason to cause localized plasticity is related to the tri-axial stress state (i.e., the stress triaxiality) induced by plastic instability at local microstructural heterogeneities (microstructural notches), namely the HAZ microstructural gradient and WM microstructural heterogeneity in the PWHT-1 and PWHT-2 material states, respectively. For the sake of completeness, it should be noted that for the weldments in PWHT-1 material state, the increased strain hardening effects were observed not only at the final fracture locations but also at the locations of the concurrent necking areas (Figure 9a). The reason for this behavior was to be expected since the c-w configuration of investigated tensile samples implied symmetrical occurrence of equivalent HAZ microstructural gradients at the both sides of welded base materials. The undeformed weldment after the renormalizing-and-tempering PWHT-2 shows equalized course of hardness values as a result of the weld microstructure homogenization (Figure 10). Moreover, the hardness profile of hydrogen-free weldment, initially subjected to the renormalizing-and-tempering PWHT-2, shows after the tensile test the localized strain hardening at the location of final failure in WM. The hydrogen-charged weldment shows the same failure location but without the occurrence of significant strain hardening at fracture location. The observed WM failures occurred likely due to the fact that the WMs represent the critical microstructural regions even after the performed homogenization treatment due to their original as-cast microstructure and thus higher impurity content and inhomogeneity compared to the BM. For comparison, Figure 11 shows four selected engineering stress-strain curves representing the overall tensile deformation behavior of the studied T92/T92 weldments in their individual material states with respect to the used PWHT conditions and hydrogen charging application. From the obtained results, it can be concluded that the tensile deformation and fracture behavior of studied T92/T92 weldments is more sensitive to the initial PWHT conditions than to the applied electrolytic hydrogenation. In spite of the occurrence of certain hydrogen embrittlement in the studied weldments, it has been shown that the thermal deterioration caused by the use of renormalizing-and-tempering PWHT-2 is much more significant. This statement is additionally supported by the fact that the final failure locations are primarily pre-determined by the used PWHT conditions, i.e., regardless of hydrogen charging application. The T92/T92 weldments after the tempering PWHT-1 were always broken in their SC-HAZs, whereas the weldments after the renormalizing-and-tempering PWHT-2 always fractured in their WMs.

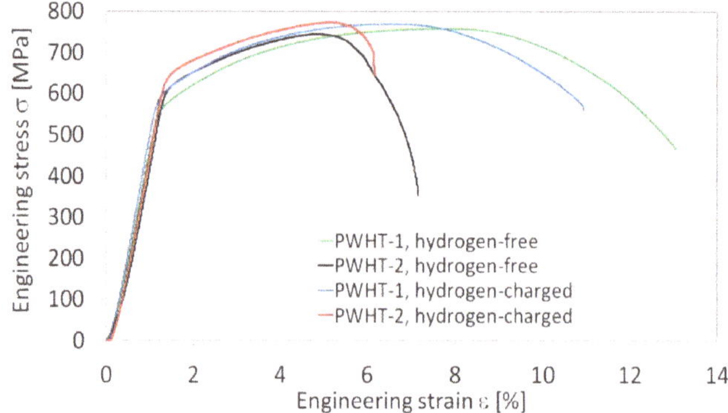

Figure 11. The effects of PWHT conditions and subsequent electrolytic hydrogenation on tensile deformation behavior of investigated T92/T92 weldments.

Figure 12 shows representative photo-macrographs of longitudinal sections of the fractured tensile samples' counterparts showing two different failure locations of the studied T92/T92 weldments in their all investigated material states.

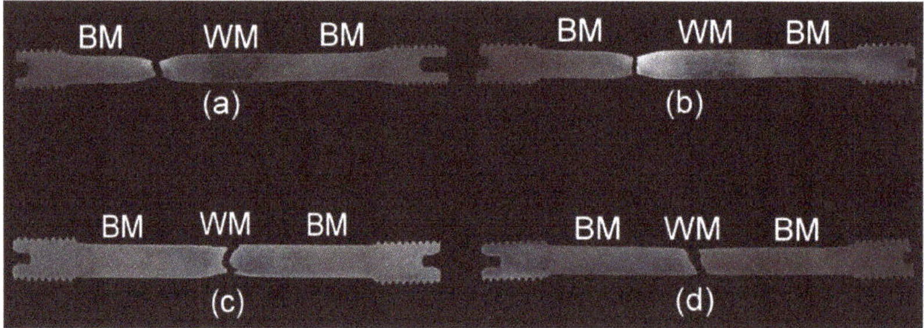

Figure 12. Photo-macrographs of fractured tensile samples' counterparts indicating various failure locations after room-temperature tensile tests of T92/T92 weldments in their individual material states: (**a**) "PWHT-1"; (**b**) "PWHT-1 + hydrogen"; (**c**) "PWHT-2"; (**d**) "PWHT-2 + hydrogen".

The detailed discussion on mutual correlation between the fractographic and microstructural characteristics of studied T92/T92 weldments will be provided within the following section.

3.3. Fractography

As already discussed, and shown in Figure 12, the room-temperature fracture behavior of studied T92/T92 weldments during tensile straining is primarily controlled by the used PWHT conditions, whereas the effect of electrolytic hydrogenation is less critical. The observation of macroscopically ductile failures (Figure 12a,b) within the SC-HAZs of both the hydrogen-free and hydrogen-charged T92/T92 weldments after the tempering PWHT-1 and tensile straining, can be related to higher deformability of the over-tempered SC-HAZs compared to the BM and WM regions. The corresponding fracture analyses for the hydrogen-free and hydrogen-charged weldments initially processed by the PWHT-1 are shown in Figures 13 and 14, respectively. Although the both hydrogen-free and hydrogen-charged weldments in PWHT-1 material state were broken in a recognizably ductile manner in their SC-HAZs (see the fracture paths in Figures 13a and 14a), the signs of hydrogen embrittlement are clearly visible for the hydrogen-charged weldment. In Figures 13a and 14a the SC-HAZ microstructures beneath final fractures are characterized by intensive deformation of grains in a direction of dominating tensile stresses. The microstructure of hydrogen-free weldment after the PWHT-1 (Figure 13a) shows randomly distributed precipitates and signs of microvoid coalescence along the elongated grain boundaries indicating the origin of the formation of ductile dimples on the fracture surface (Figure 13b). Whereas the fracture surface of hydrogen-free T92/T92 weldment in PWHT-1 material state shows completely ductile dimple fracture (Figure 13b), the fracture surface of the hydrogen-charged weldment shows quasi-cleavage areas with typical "fish-eye" morphology due to the local hydrogen-embrittlement beside ductile dimple tearing areas (Figure 14b). The hydrogenated sample after the PWHT-1 shows in its microstructure beneath the fracture surface hydrogen-assisted cracks and even some grain fragmentation (Figure 14a). The "fish-eye" fractographic object on the fracture surface (Figure 14b) represents the hydrogen embrittled quasi-cleavage zone created radially along the central inclusion.

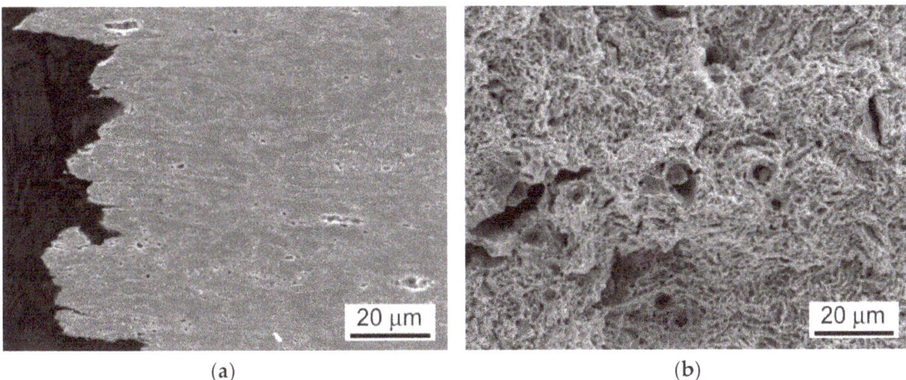

(a)   (b)

Figure 13. Fracture analysis of hydrogen-free T92/T92 weldment initially processed by the PWHT-1 and subsequently ruptured in SC-HAZ after the room-temperature tensile test: (**a**) fracture path and microstructure beneath the fracture; (**b**) fracture surface showing pure ductile dimple tearing.

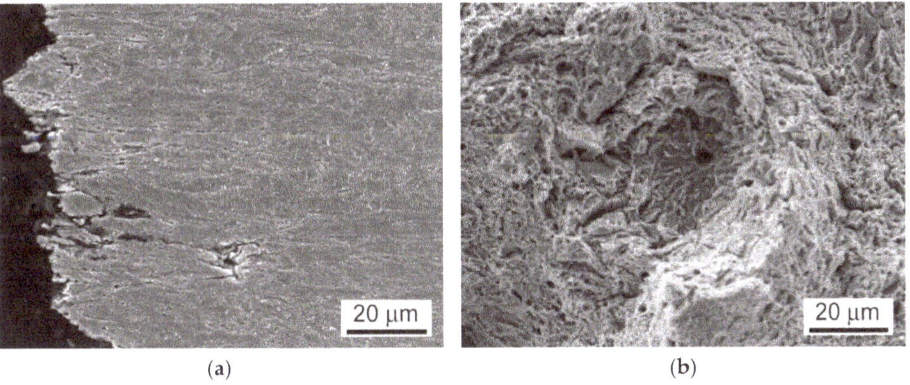

(a)   (b)

Figure 14. Fracture analysis of hydrogen-charged T92/T92 weldment initially processed by the PWHT-1 and subsequently ruptured in SC-HAZ after the room-temperature tensile test: (**a**) fracture path and microstructure beneath the fracture; (**b**) fracture surface showing mixed fracture consisting of both transgranular quasi-cleavage with a "fish-eye" morphology and ductile dimple tearing.

On the other hand, the observation of rather brittle failures (Figure 12c,d) within the WMs of both the hydrogen-free and hydrogen-charged T92/T92 weldments after the renormalizing-and-tempering PWHT-2 can be related to their lower deformability. It appears as a result of thermal embrittlement and possibly remaining microstructural heterogeneities of the original as-cast WMs microstructures (e.g., dendritic segregation) due to insufficient homogenization (i.e., short time for the alloying elements redistribution during performed reaustenitization, see Figure 1). Thus, the WMs represent the most heterogeneous regions within the renormalized-and-tempered weldments. This finding indicates the reason for their high propensity for the localization of plastic deformation and final failure occurrence during tensile straining. The corresponding fracture analyses for the hydrogen-free and hydrogen-charged weldments initially processed by the PWHT-2 are shown in Figures 15 and 16, respectively.

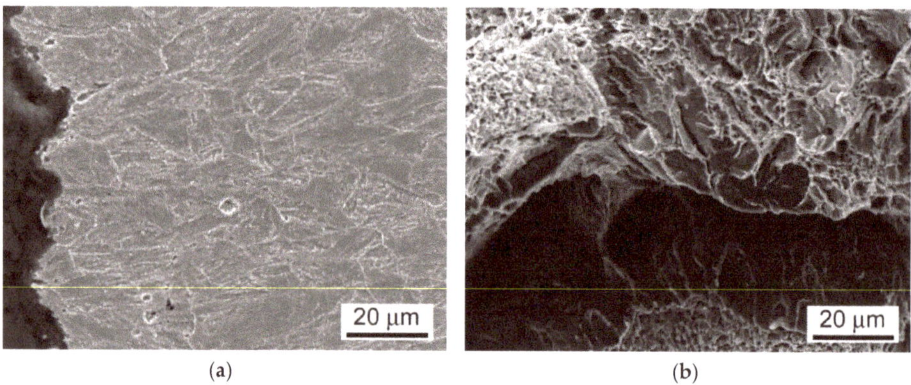

Figure 15. Fracture analysis of hydrogen-free T92/T92 weldment initially processed by the PWHT-2 and subsequently ruptured in WM after the room-temperature tensile test: (**a**) fracture path and microstructure beneath the fracture; (**b**) fracture surface showing mixed fracture consisting of both ductile dimple tearing and transgranular quasi-cleavage.

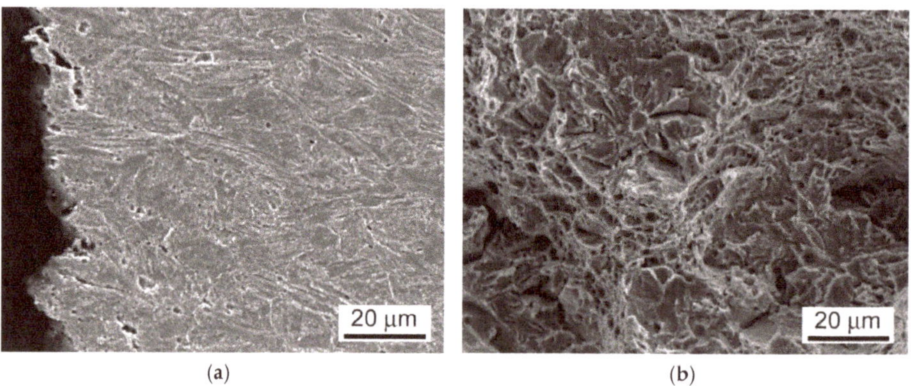

Figure 16. Fracture analysis of hydrogen-charged T92/T92 weldment initially processed by the PWHT-2 and subsequently ruptured in WM after the room-temperature tensile test: (**a**) fracture path and microstructure beneath the fracture; (**b**) fracture surface showing mixed fracture consisting of transgranular quasi-cleavage with a "fish-eye" morphology besides ductile dimple ridges and some inter-lath decohesion areas.

The fracture paths in Figures 15a and 16a indicate lower ductility failures for both the hydrogen-free and hydrogen-charged T92/T92 weldments after the renormalizing-and-tempering PWHT-2 compared to the weldments after the tempering PWHT-1 (Figures 13a and 14a). The clearer differences between the fracture mechanisms related to the hydrogen-free and hydrogen-charged weldments after the PWHT-2 are distinguishable on the fracture surfaces in Figures 15b and 16b. Whereas the fracture surface of hydrogen-free weldment exhibits mixed fracture mechanisms including the ductile dimple tearing and transgranular quasi-cleavage (Figure 15b), the fracture surface of hydrogenated weldment shows even more complex fracture mechanisms involving transgranular quasi-cleavage with a "fish-eye" morphology besides ductile dimple tearing and some inter-lath decohesion (Figure 16b). The reason for the observed fracture morphology changes can be explained by considering local microstructural characteristics beneath the fracture surfaces. In Figures 15a and 16a the WM microstructures beneath final fractures are characterized by polygonal grains and sub-grains almost continuously decorated with coarsened secondary phase precipitates indicating thermal embrittlement induced by PWHT-2. That is why the both hydrogen-free and hydrogen-charged samples show mixed fractures consisting of both ductile dimple tearing and brittle transgranular quasi-cleavage (Figures 15b and 16b). The reason for the additional occurrence of "fish-eye" fracture morphology on the fracture surface of the hydrogenated sample treated by PWHT-2 (Figure 16b) is basically the same as in the case of hydrogenated sample after the PWHT-1. However, the higher level of thermal embrittlement after the PWHT-2 was at the same time the reason for higher grade of hydrogen embrittlement due to the existence of more nucleation sites for the creation of "fish-eye" fractographic objects.

As already shown in previous section, the hardness profile of undeformed c-w sample of T92/T92 weldment after the PWHT-2 (Figure 10) showed somewhat lower hardness in WM compared to the rest BM portions. This observation may also be related to lower alloying of WM compared to BM. Thus, the preferential deformation of softer WMs during room-temperature tensile deformation of the welds after the PWHT-2 gave rise to their higher propensity for localization of plastic deformation and final failure occurrence. It can be concluded that the use of renormalizing-and-tempering PWHT-2 was found to be rather inconvenient for improving the microstructure and mechanical properties of the investigated T92/T92 weldments. However, based on the results obtained from the cross-weld tensile tests, the hydrogen embrittlement susceptibility of the investigated weldments was rather low for both the applied PWHT conditions.

4. Conclusions

In present study the T92/T92 weldments in two different initial PWHT states (i.e., in either conventionally tempered PWHT-1 state or in renormalized-and-tempered PWHT-2 state) were investigated in terms of the electrolytic hydrogenation effect on their room-temperature tensile properties and fracture behavior. Here are the main conclusions:

- The observed microstructural variations of studied weldments induced by different initial PWHT conditions had crucial effects on their resulting tensile deformation and fracture behavior in both the hydrogen-free and hydrogen-charged conditions. The weldments after the tempering PWHT-1 exhibited a typical HAZ microstructural gradient consisting of the CG-HAZ, FG-HAZ, IC-HAZ, and SC-HAZ, whereas the weldments after the renormalizing-and-tempering PWHT-2 showed quite homogenized microstructure of former HAZ. In addition, the regions of WM and former CG-HAZ showed certain microstructural refinement compared to the rest of the BM.
- The effects of both the PWHT conditions applied and electrolytic hydrogenation on the resulting strength properties, i.e., the YS and UTS values of studied weldments, were rather insignificant. A small increase of the measured strength properties has been observed after the renormalizing-and-tempering PWHT-2 compared to the tempering PWHT-1, which was reasonably attributed to the overall suppression of the former HAZ microstructural gradient with its soft microstructural sub-regions as indicated by c-w hardness measurements. As expected, some small hydrogen-induced hardening effects were observed for the weldments in both the PWHT-1 and PWHT-2 states.
- The used renormalizing-and-tempering PWHT-2 of studied weldment resulted in a significant decrease of its plastic properties, i.e., the EL and RA values, compared to those of the weldment after the conventional tempering PWHT-1. Thus, the use of renormalizing-and-tempering PWHT-2 was found to be quite inappropriate for improving the microstructure and mechanical properties of the investigated T92/T92 weldments. The application of electrolytic hydrogenation of the

studied weldments in their both PWHT conditions led to additional detrimental effects of their plastic properties. Although the results indicated recognizably higher hydrogen embrittlement susceptibility for the renormalized-and-tempered weldments compared to the conventionally tempered ones, it can be concluded that all studied weldments show sufficient resistance against hydrogen embrittlement in conditions of the present investigation.

Author Contributions: Conceptualization, L.Č. and L.F.; methodology, L.Č., M.D., V.H. and R.D.; formal analysis, L.Č., L.F., and I.D.; investigation, L.Č., L.F., M.D.; data curation, L.Č. and V.H.; writing—original draft preparation, L.F.; writing—review and editing, L.F. and I.D.; visualization, L.Č.; supervision, L.F. and I.D.; project administration, L.F.; funding acquisition, L.F. All authors have read and agreed to the published version of the manuscript.

Funding: This research was funded by "Vedecká Grantová Agentúra MŠVVaŠ SR a SAV" (Grant No.: VEGA 2/0062/19).

Conflicts of Interest: The authors declare no conflict of interest.

References

1. Parker, J.; Siefert, J. Manufacture and performance of welds in creep strength enhanced ferritic steels. *Materials* **2019**, *12*, 2257. [CrossRef] [PubMed]
2. El-Desoky, O.E.; Abd El-Azim, M.E.; ElKossy, M.R. Analysis of creep behavior of welded joints of P91 steel at 600 °C. *Int. J. Press. Vessels Pip.* **2019**, *171*, 145–152. [CrossRef]
3. Qiao, S.; Wei, Y.; Xu, H.; Cui, H.; Lu, F. The evolution behavior of second phases during long-term creep rupture process for modified 9Cr-1.5Mo-1Co steel welded joint. *Mater. Charact.* **2019**, *151*, 318–331. [CrossRef]
4. Wei, Y.; Qiao, S.; Lu, F.; Liu, W. Failure transition mechanism in creep rupture of modified casting 9Cr-1.5Mo-1Co welded joint. *Mater. Des.* **2016**, *975*, 268–278. [CrossRef]
5. Wu, T.-J.; Liao, C.-C.; Chen, T.-C.; Shiue, R.-K.; Tsay, L.-W. Microstructural evolution and short-term creep rupture of the simulated HAZ in T92 steel normalized at different temperatures. *Metals* **2019**, *9*, 1310. [CrossRef]
6. Saini, N.; Mulik, R.S.; Mahapatra, M.M. Influence of filler metals and PWHT regime on the microstructure and mechanical property relationships of CSEF steels dissimilar welded joints. *Int. J. Press. Vessels Pip.* **2019**, *170*, 1–9. [CrossRef]
7. Pandey, C.; Mahapatra, M.M.; Kumar, P.; Thakre, J.G.; Saini, N. Role of evolving microstructure on the mechanical behaviour of P92 steel welded joint in as-welded and post weld heat treated state. *J. Mater. Process. Technol.* **2019**, *263*, 241–255. [CrossRef]
8. Milović, L.; Vuherer, T.; Vrhovac, M.; Stankovic, M. Microstructures and mechanical properties of creep resistant steel for application at elevated temperatures. *Mater. Des.* **2013**, *46*, 660–667. [CrossRef]
9. Schüller, H.J.; Hagn, L.; Woitscheck, A. Risse im Schweißnahtbereich von Formstücken aus Heißdampfleitungen. *Werkstoffuntersuchungen der Maschinenschaden* **1974**, *47*, 1–13.
10. Cerjak, H.; Mayr, P. Creep strength of welded joints of ferritic steels. In *Creep-Resistant Steels*, 1st ed.; Abe, F., Kern, T.-U., Viswanathan, R., Eds.; Woodhead publishing Ltd.: Cambridge, UK, 2008; pp. 472–498.
11. Chang, J.C.; Kim, B.S.; Heo, N.H. Stress relief cracking on the weld of T/P 23 Steel. *Procedia Eng.* **2011**, *10*, 734–739. [CrossRef]
12. Wang, X.; Li, Y.; Li, H.; Lin, S.; Ren, Y. Effect of long-term aging on the microstructure and mechanical properties of T23 steel weld metal without post-weld heat treatment. *J. Mater. Process. Technol.* **2018**, *252*, 618–627. [CrossRef]
13. Mohyla, P.; Foldyna, V. Improvement of reliability and creep resistance in advanced low-alloy steels. *Mater. Sci. Eng. A* **2009**, *510–511*, 234–237. [CrossRef]
14. Brett, S.J. Type IIIa cracking in $\frac{1}{2}$CrMoV steam pipework systems. *Sci. Technol. Weld. Join.* **2004**, *9*, 41–45. [CrossRef]
15. Blach, J.; Falat, L.; Ševc, P. The influence of hydrogen charging on the notch tensile properties and fracture behaviour of dissimilar weld joints of advanced Cr–Mo–V and Cr–Ni–Mo creep-resistant steels. *Eng. Fail. Anal.* **2011**, *18*, 485–491. [CrossRef]

16. Bruycker, E.D.; Huysmans, S.; Vanderlinden, F. Investigation of the hydrogen embrittlement susceptibility of T24 boiler tubing in the context of stress corrosion cracking of its welds. *Procedia Struct. Integr.* **2018**, *13*, 226–231. [CrossRef]

17. Xue, W.; Pan, Q.; Ren, Y.; Shang, W.; Zeng, H.; Liu, H. Microstructure and type IV cracking behavior of HAZ in P92 steel weldment. *Mater. Sci. Eng. A* **2012**, *552*, 493–501. [CrossRef]

18. Zhao, L.; Jing, H.; Xu, L.; Han, Y.; Xiu, J. Experimental study on creep damage evolution process of Type IV cracking in 9Cr–0.5Mo–1.8W–VNb steel welded joint. *Eng. Fail. Anal.* **2012**, *19*, 22–31. [CrossRef]

19. Albert, S.K.; Matsui, M.; Watanabe, T.; Hongo, H.; Kubo, K.; Tabuchi, M. Variation in the type IV cracking behaviour of a high Cr steel weld with post weld heat treatment. *Int. J. Press. Vessel. Pip.* **2003**, *80*, 405–413. [CrossRef]

20. Pandey, C.; Mahapatra, M.M.; Kumar, P.; Kumar, S.; Sirohi, S. Effect of post weld heat treatments on microstructure evolution and type IV cracking behavior of the P91 steel welds joint. *J. Mater. Process. Technol.* **2019**, *266*, 140–154. [CrossRef]

21. Parker, J.D.; Stratford, G.C. The high-temperature performance of nickel-based transition joints—II. Fracture behaviour. *Mater. Sci. Eng. A* **2001**, *299*, 179. [CrossRef]

22. Falat, L.; Ševc, P.; Čiripová, L. Creep cracking of similar and dissimilar T92 steel weldments. *Powder Metall. Prog.* **2015**, *15*, 130–137.

23. Francis, J.A.; Mazur, W.; Bhadeshia, H.K.D.H. Review Type IV cracking in ferritic power plant steels. *Mater. Sci. Technol.* **2006**, *22*, 1387–1395. [CrossRef]

24. Silwal, B.; Li, L.; Deceuster, A.; Griffiths, B. Effect of postweld heat treatment on the toughness of heat-affected zone for grade 91 steel. *Weld. J.* **2013**, *92*, 80.

25. Rieth, M.; Rey, J. Specific welds for test blanket modules. *J. Nucl. Mater.* **2009**, *386*, 471–474. [CrossRef]

26. Widak, V.; Dafferner, B.; Heger, S.; Rieth, M. Investigations of dissimilar welds of the high temperature steels P91 and PM2000. *Fusion Eng. Des.* **2013**, *88*, 2539–2542. [CrossRef]

27. Falat, L.; Výrostková, A.; Svoboda, M.; Milkovič, O. The influence of PWHT regime on microstructure and creep rupture behaviour of dissimilar T92/TP316H ferritic/austenitic welded joints with Ni-based filler metal. *Kovove. Mater.* **2011**, *49*, 417–426. [CrossRef]

28. Falat, L.; Čiripová, L.; Kepič, J.; Buršík, J.; Podstranská, I. Correlation between microstructure and creep performance of martensitic/austenitic transition weldment in dependence of its post-weld heat treatment. *Eng. Fail. Anal.* **2014**, *40*, 141–152. [CrossRef]

29. Falat, L.; Kepič, J.; Čiripová, L.; Ševc, P.; Dlouhý, I. The effects of postweld heat treatment and isothermal aging on T92 steel heat-affected zone mechanical properties of T92/TP316H dissimilar weldments. *J. Mater. Res.* **2016**, *31*, 1532–1543. [CrossRef]

30. Ševc, P.; Falat, L.; Čiripová, L.; Džupon, M.; Vojtko, M. The Effects of electrochemical hydrogen charging on room-temperature tensile properties of T92/TP316H dissimilar weldments in quenched-and-tempered and thermally-aged conditions. *Metals* **2019**, *9*, 864. [CrossRef]

31. Čiripová, L.; Falat, L.; Ševc, P.; Vojtko, M.; Džupon, M. Ageing effects on room-temperature tensile properties and fracture behavior of quenched and tempered T92/TP316H dissimilar welded joints with Ni-based weld metal. *Metals* **2018**, *8*, 791. [CrossRef]

32. Blach, J.; Falat, L. The influence of thermal exposure and hydrogen charging on the notch tensile properties and fracture behaviour of dissimilar T91/TP316H weldments. *High Temp. Mater. Process.* **2014**, *33*, 329–337. [CrossRef]

33. Nguyen, L.T.H.; Hwang, J.-S.; Kim, M.-S.; Kim, J.-H.; Kim, S.-K.; Lee, J.-M. Charpy impact properties of hydrogen-exposed 316L sainless steel at ambient and cryogenic temperatures. *Metals* **2019**, *9*, 625. [CrossRef]

34. Jia, H.; Zhang, X.; Xu, J.; Sun, Y.; Li, J. Effect of hydrogen content and strain rate on hydrogen-induced delay cracking for hot-stamped steel. *Metals* **2019**, *9*, 798. [CrossRef]

35. Dunne, D.P.; Hejazi, D.; Saleh, A.A.; Haq, A.J.; Calka, A.; Pereloma, E.V. Investigation of the effect of electrolytic hydrogen charging of X70 steel: I. The effect of microstructure on hydrogen-induced cold cracking and blistering. *Int. J. Hydrog. Energy* **2016**, *41*, 12411–12423. [CrossRef]

36. Yin, C.; Chen, J.; Ye, D.; Xu, Z.; Ge, J.; Zhou, H. Hydrogen concentration distribution in 2.25cr-1mo-0.25v steel under the electrochemical hydrogen charging and its influence on the mechanical properties. *Materials* **2020**, *13*, 2263. [CrossRef] [PubMed]

37. Sklenička, V.; Kuchařová, K.; Svobodová, M.; Kvapilová, M.; Král, P.; Horváth, L. Creep properties in similar weld joint of a thick-walled P92 steel pipe. *Mater. Charact.* **2016**, *119*, 1–12. [CrossRef]
38. Cao, J.; Gong, Y.; Zhu, K.; Yang, Z.-G.; Luo, X.-M.; Gu, F.-M. Microstructure and mechanical properties of dissimilar materials joints between T92 martensitic and S304H austenitic steels. *Mater. Des.* **2011**, *32*, 2763–2770. [CrossRef]
39. Pandey, C.; Mahapatra, M.M.; Kumar, P.; Saini, N. Dissimilar joining of CSEF steels using autogenous tungsten-inert gas welding and gas tungsten arc welding and their effect on δ-ferrite evolution and mechanical properties. *J. Manuf. Process.* **2018**, *31*, 247–259. [CrossRef]
40. Barbadikar, D.R.; Deshmukh, G.S.; Maddi, L.; Laha, K.; Parameswaran, P.; Ballal, A.R.; Peshwe, D.R.; Paretkar, R.K.; Nandagopal, M.; Mathew, M.D. Effect of normalizing and tempering temperatures on microstructure and mechanical properties of P92 steel. *Int. J. Press. Vessel. Pip.* **2015**, *132–133*, 97–105. [CrossRef]
41. Kral, P.; Dvorak, J.; Sklenicka, V.; Masuda, T.; Horita, Z.; Kucharova, K.; Kvapilova, M.; Svobodova, M. The effect of ultrafine-grained microstructure on creep behaviour of 9% Cr steel. *Materials* **2018**, *11*, 787. [CrossRef]

 *materials*

Article

Can Sub-Zero Treatment at −75 °C Bring Any Benefits to Tools Manufacturing?

Martin Kusý [1], Lýdia Rízeková-Trnková [1], Jozef Krajčovič [1], Ivo Dlouhý [2] and Peter Jurči [1,*]

[1] Institute of Materials Science, Faculty of Materials Science and Technology in Trnava, Paulínská 16, 917 24 Trnava, Slovakia; martin.kusy@stuba.sk (M.K.); lydia.trnkova@stuba.sk (L.R.-T.); jozef_krajcovic@stuba.sk (J.K.)

[2] Institute of Physics of Materials, CEITEC-IPM, Czech Academy of Sciences, Zizkova 22, 61662 Brno, Czech Republic; idlouhy@ipm.cz

* Correspondence: p.jurci@seznam.cz

Received: 22 October 2019; Accepted: 20 November 2019; Published: 21 November 2019

Abstract: Vanadis 6 ledeburitic tool steel was subjected to sub-zero treatment at −75 °C for different durations, and for different subsequent tempering regimes. The impact of these treatments on the microstructure, hardness variations, and toughness characteristics of the steel was investigated. The obtained results infer that the retained austenite amount was reduced to one fourth by sub-zero treatment (SZT), and the population density of add-on carbides was increased by factor of three to seven, depending on the duration of SZT. Tempering always reduced the population density of these particles. A hardness increased by 30–60 HV10 was recorded after sub-zero treatment but tempering to the secondary hardness peak induced much more significant hardness decrease than what was established in conventionally quenched steel. The flexural strength was not negatively influenced by sub-zero treatment at −75 °C while the fracture toughness tests gave worse values of this quantity, except the case of steel tempered to the secondary hardness peak.

Keywords: vanadis 6 steel; sub-zero treatment at −75 °C; microstructure; hardness; fracture toughness

1. Introduction

In conventional heat treatment of tools made of high-alloyed Cr and Cr–V ledeburitic steels, the material is gradually heated up to the austenitizing temperature (recommended by the steel manufacturers), held there for prescribed time, and cooled down rapidly to the room temperature. Afterwards the steels should be subjected to tempering immediately, to prevent the retained austenite stabilization, and to induce transformations leading to the achievement of final properties of tools or components.

Ledeburitic steels are commonly used as tool materials for cold work applications. In order to generate high abrasive wear resistance, they contain a high amount of carbide phases embedded in a metallic matrix. On the other hand, these carbides together with high overall steel hardness deteriorate the material resistance against crack initiation/propagation, expressed by the fracture toughness K_{IC}. Also, the flexural strength of ledeburitic steels, which is often also taken as a measure of resistance against crack initiation for brittle materials, manifests relatively low values. Moreover, conventionally produced ledeburitic steels are cast and afterwards hot formed. As a consequence, they contain band-like carbides, thus they suffer from anisotropy of key mechanical properties [1].

Sub-zero treatment is defined as a supplementary process to the conventional heat treatment. Unlike conventional heat treatment (CHT), it is a process where the tools or components are immersed into suitable cryoprocessing medium, stored there for pre-determined time (usually in tens of hours), and re-heated to the room temperature. Research works conducted on application of this kind of treatment have shown that sub-zero treatments provide extra benefits to the tooling industry like increased hardness [2,3], better wear performance [3,4], and improved dimensional stability of products [5].

According to recent studies, the following crucial microstructural changes are responsible for these benefits:

(i)   Considerably reduced retained austenite (γ_R) amount [2,6–9].
(ii) Refinement of the martensite along with an enhanced number of crystal defects such as dislocations and twins inside martensitic domains [7,9,10].
(iii) Enhancement of number and population density of small globular carbides (SGCs) [2,3,7,8,11,12].
(iv) Acceleration of the precipitation rate of nano-sized transient carbides [11–15].

However, the impact of sub-zero treatments on the toughness characteristics (fracture toughness, flexural strength, and impact toughness) is controversial as Table 1 illustrates.

Table 1. Toughness and fracture toughness of differently sub-zero treated ledeburitic steels—an overview of the obtained results to date.

Steel	Treatment	Quantity	Description	Reference
AISI D2	−70, −100, −130, and −196 °C, tempering at 200 °C	Impact toughness	Considerable reduction, extent of reduction depends on both the austenitizing temperature (better toughness is achieved at lower austenitizing temperature) and SZT temperature (−70 °C produces the lowest values)	[16]
AISI D2	−196 °C/16 h, tempering at 170 or 450 °C	Impact toughness	Considerable reduction, degree of reduction is much higher than the improvement of hardness	[17]
AISI D2	−196 °C/20 h	Impact toughness	Reduction after low-temperature tempering but improvement after tempering to the secondary hardness	[18]
AISI D2	−196 °C/36 h, −75 °C/5 min −125 °C/5 min, tempering at 210 °C	Fracture toughness	Significant reduction after SZT at either −75 or −125 °C, moderate reduction after SZT at −196 °C	[19]
Vanadis 6	−196 °C/4 or 10 h, −90 °C/4 h, tempering twice at 530 °C	Flexural strength, fracture toughness	Marginal effect on flexural strength but slightly positive effect on fracture toughness	[20]
Vanadis 6	−196 °C/17 h, tempering 170–530 °C	Flexural strength, fracture toughness	Marginal effect on flexural strength in low-temperature Tempering range, improvement in normal secondary hardening temperature range. Worsening of fracture toughness except tempering to the secondary hardness.	[21]
Vanadis 6	−140 °C/17 h, tempering at 170–530 °C	Flexural strength, fracture toughness	Slight improvement of flexural strength at low tempering temperatures but almost no effect after tempering at 450 °C and above. Improvement of fracture toughness after high temperature tempering but slight deterioration after tempering within the 170–450 °C range	[22]
AISI D3	−196 °C/12, 24 or 36 h, tempering at 150 °C	Impact toughness	Significant reduction, the reduction is more pronounced with increasing the duration of SZT	[23]

The question of an optimal regime of sub-zero treatments is still under debate. In the "pioneer age" of this technique it was believed, within the professional community, that the benefits of sub-zero treatments are based only on the reduction of retained austenite amount. Therefore, the temperatures of around −75 °C were widely used in laboratory and industrial practice. Lower temperatures were not accepted for the treatments since their use often led to premature failure of tools, due to thermal shocks associated with the use of very low temperatures. Treatment at the boiling temperature of liquid nitrogen (−196 °C) was introduced into industrial practice only much later, when the devices enabled to carry out well controlled cooling down to such low temperature.

Other sub-zero treatment temperatures were suggested only by very limited number researchers. Reitz and Pendray and Gavriljuk et al., for instance, suggested the temperature of −140 °C [24,25]. Recent studies dealing with thorough analysis of microstructure and toughness of the Vanadis 6 steel treated in this way gave very promising results [22,26]. Alternatively, there were attempts and/or suggestions with the use of the temperature of boiling helium (−269 °C) [13,27]. However, the treatment temperature of −75 °C also deserves attention since one can expect that the phenomena being responsible for abovementioned ameliorations in properties would proceed faster at −75 °C than at lower temperatures. Also, practical experiences indicated that the extent of "extra" wear performance (or other property), which can be gained by the use of SZT at −196 °C (as compared with treatments at −75 °C) depends on the material chemistry. For instance, the wear performance of AISI D2 steel was improved by a factor of 2.59 by the treatment in liquid nitrogen (as compared with treatment at −75 °C), while only an improvement by a factor of 1.39 was recorded for CPM 10-V steel (steel with high vanadium content) [24].

The current paper is thus focused to an in-depth description of the results obtained with the sub-zero treatments of Vanadis 6 steel at the temperature of −75 °C, and to their careful discussion. Microstructural changes are presented, and they are related to the hardness, flexural strength, and fracture toughness of examined steel. The obtained results are also compared to what was obtained by treatments at temperatures −140, −196, and −269 °C, respectively.

2. Experimental

The powder metallurgy (PM) tool steel, Vanadis 6, with nominal composition (in wt%) 2.1 C, 1 Si, 0.4 Mn, 6.8 Cr, 1.5 Mo, 5.4 V, and Fe as a balance was selected for the examinations. Due to the PM technique used for the steel manufacturing the material is free of macrosegregations and also of carbide bands, and manifests high degree of isotropy. This makes it possible to disregard the orientation in sample manufacturing. The initial microstructure was soft-annealed, with a hardness of 272 HV10.

The specimens were machined to a net-shape and subjected to the heat treatment schedules. Conventional heat treatment consisted of gradual heating up to the austenitizing temperature of 1050 °C, holding at that temperature for 30 minutes to enable the dissolution of carbides and austenite homogenization, which was followed by room temperature quenching by using cold nitrogen gas. Then, one set of specimens was separated and subjected instantly to tempering treatments. The other specimens were moved to cryogenic system and subjected to sub-zero treatments at the temperature of −75 °C, and for different durations prior to tempering.

In sub-zero treatments, room temperature quenched specimens were cooled down to −75 °C, at a controlled cooling rate of 1 °C/min, stored at the lowest temperature for predetermined duration (4, 10, 17, 24, or 48 h), and re-heated by a heating rate of 1 °C/min to the room temperature. Immediately after the specimens reached the room temperature, they were moved to the tempering furnace where they were tempered in an atmosphere of pure technical nitrogen. Tempering consisted of two cycles (2 + 2 h), at temperatures in the range 170–530 °C. However, the tempering temperature of 600 °C was added to the experiments with flexural strength, in order to verify the behavior of this mechanical quantity at tempering temperature that is located beyond the secondary hardness peak.

A NETZSCH DIL 402 C dilatometer (Netzsch, Selb, Germany) was used within the temperature range of 20 down to −150 °C in order to estimate the martensite finish temperature of the Vanadis 6 steel austenitized at given austenitizing conditions.

The specimens for microstructural examinations were prepared using standard metallographic grinding by a set of abrasive papers (with a grit of 180, 320, 600, and 1200) and polishing (by using the 9, 3, and 1 μm diamond suspensions). Then, the specimens were etched with Villella-Bain reagent (5 ml hydrochloric acid, 1 g picric acid in 100 ml of ethanol) for 10 s. For the material microstructure examinations, a JEOL JSM 7600 F scanning electron microscope (SEM, Jeol Ltd., Tokyo, Japan), operating at a 15 kV acceleration voltage was used. Microstructural examinations were coupled with energy dispersive spectrometry (EDS), by using an EDS detector (Oxford Instruments, plc.,

High Wycombe, UK). SEM micrographs were acquired in a combined 50:50 detection of secondary electrons and backscattered electrons (BE). The reason was that the material contains eutectic carbides (ECs), secondary carbides (SCs), and small globular carbides after application of abovementioned heat treatment schedules. It has been proven recently that the ECs are represented by vanadium-rich (more than 50 wt%V) MC particles, while the SCs are the M_7C_3 particles (with a high percentage of chromium), and that SGCs were determined to be alloyed cementite [7]. This made it possible to clearly differentiate between the ECs on the one side, and the SCs (and the SGCs) on the other side, by strong differences in BE yield. Unfortunately, this categorization fails in differentiation between SCs and SGCs as their BE yield is very similar. Therefore, the classification based on the particle size was adopted to distinguish the ECs and the SGCs; the carbides finer than 0.5 µm were considered as the SGCs while the coarser ones as SCs.

Determination of population density of carbide particles (categorized as above-described) was carried out on twenty-five randomly acquired SEM micrographs for each specimen. Standard magnification for acquisition of SEM images was 3000×. For better identification of SGCs additional SEM micrographs, at a magnification of 7500× were recorded because some of these particles had a size well below 100 nm. The acquisition of SEM images was coupled with EDS mapping of chromium and vanadium, in order to clearly differentiate between the carbides, as above mentioned. The mean values and the standard deviations of the obtained data were calculated.

The phase constitution of differently heat-treated specimens was determined by using X-ray diffraction (XRD, Philips Analytical B.V., Almelo, The Netherlands) technique. A Phillips PW 1710 diffractometer with filtered $Co_{K\alpha}$ characteristic radiation was used for this purpose. The diffracted radiation was registered within a two-theta angle range of 30–127 deg. The retained austenite amounts were determined following the appropriate ASTM E975-13 standard [28], taking the characteristic peaks of both the martensite (α') and the austenite (γ), namely (200)α', (200)γ, (211)α', and (220)γ into the consideration. As reported recently [2,7], however, these peaks are often superimposed by the characteristic peaks of carbides, which might influence the accuracy of the obtained results negatively. Therefore, the analyses were coupled with Rietveld refinement of the obtained X-ray spectra before computing the retained austenite amounts.

Hardness of differently heat-treated specimens was measured by using a Vickers indentation technique, at a load of 98.1 N (HV 10), following the appropriate Czech standard [29]. A ZWICK 3212 hardness tester (Zwick-Roell, Ulm, Germany) was used. The distance between two adjacent indents was kept minimum 5 mm, and the dwell time used for each indent was 15 s. Ten measurements were done for each specimen. Then, both the mean values and standard deviations of measured values were calculated.

10 by 10 by 100 mm bar specimens were used for flexural strength determination. Prior to measurements the specimens were polished to a final surface roughness, R_a, between 0.05 and 0.07 µm. This step is very important to obtain reliable results because it is known that the surface finish plays an important role in this type of measurement, and differences in surface finish (in microns of R_a) can influence the obtained flexural strength values within the range of hundreds MPa [30]. An Instron 8862 test device (Instron, Norwood, MA, USA) was used. Specimens were tested in three-point bending configuration, at a loading rate of 1 mm/min, until the moment of fracture. The distance between loading roller supports was 80 mm. The flexural strength R was calculated according to equation (following the appropriate Czech standard [31])

$$R = \frac{3FL}{2bh^2} \text{ [MPa]} \tag{1}$$

where F represents fracture force (maximum load on the load–deflection trace), L is distance of roller supports in three-point bending, b is the specimen thickness, and h is the specimen height (dimension in the direction of the acting load). In addition, the total work of fracture, W_{of}, and the plastic part of the work of fracture, W_{pl}, were evaluated from the corresponding area below the measured load-deflection (load displacement) curve.

Reliable evaluation of fracture toughness of materials by plane strain fracture toughness requires prior fatigue pre-cracking of the specimens in order to achieve a sharp and reproducible crack tip geometry for testing. Therefore, pre-cracked specimens with 10 by 10 by 55 mm dimensions were used for fracture toughness determination in the current work. Pre-cracking the samples was carried out after the heat treatment. A resonance frequency machine Cracktronic 8024 (Russenberger Prüfmaschinen AG, Neuhausen am Rheinfall, Switzerland) was used for this purpose, and the specimens were loaded in four-point bending configuration. The crack development was monitored on both sides of the sample using a digital long-distance microscope. Both the pre-crack preparation and the testing were carried out at room temperature following the ISO 12137 standard [32]. In testing, the specimens were loaded in three-point bending with a roller span of 40 mm, and at a loading rate of 0.1 mm/min. An Instron 8862 machine (Instron, Norwood, MA, USA) was used. Specimen deflection was measured by means of an inductive transducer integrated directly into the loading axis. Five samples were tested for each investigated heat treatment (SZT, tempering) condition.

Fracture surfaces were analyzed by using the scanning electron microscope JEOL JSM 7600 F. SEM micrographs were acquired in the secondary electrons detection regime, at different magnifications enabling to study the micro-morphology of fracture surfaces. Particular attention has been paid to the role of carbide particles in the fracture propagation.

3. Results and Discussion

The change in the length of the Vanadis 6 specimen after quenching from 1050 °C down to 20 °C and immediate moving to the dilatometer, where the material was cooled down from 20 °C to −110 °C is presented in Figure 1.

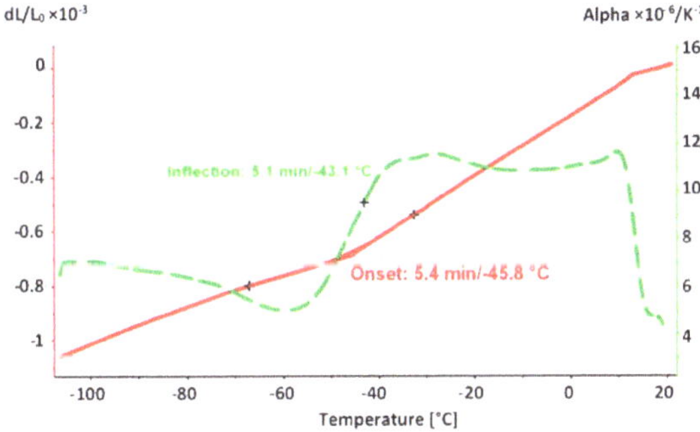

Figure 1. Dilatometry of Vanadis 6 steel after austenitizing at 1050 °C for 30 min, quenching to 20 °C, and cooling from 20 °C down to −110 °C with a cooling rate of 10 K min⁻¹.

The onset at the relative length curve is located at −45.8 °C. The inflection at the linear expansion coefficient curve is at −43.1 °C. From these two values it can be summarized that the critical M_f temperature of the Vanadis 6 steel lies at approximately −45 °C, it is thus higher than the selected sub-zero treatment temperature.

SEM micrographs, Figure 2, show the microstructure of Vanadis 6 steel in the conventionally heat-treated state (a), and after subsequent SZT at −75 °C for different durations (b–f). The matrix and different carbide particles are the main microstructural constituents of the steel, irrespectively to the heat treatment schedule used. The matrix is mainly martensitic, with certain amount of retained austenite. The retained austenite is located in-between the martensitic laths as SEM micrograph in Figure 2a illustrates. The carbides particles are the eutectic carbides, the secondary carbides, and small globular carbides. The number and population density of both the ECs and the SCs are nearly constant within the range of heat treatment schedules used. Alternatively, the number and population density of SGCs change with the duration of SZT; they increase substantially up to the 10 h duration of SZT, Figure 2b,c, where the maximum population density of 233×10^3 mm^{-2} was achieved (Table 2), and then decrease, Figure 2d–f.

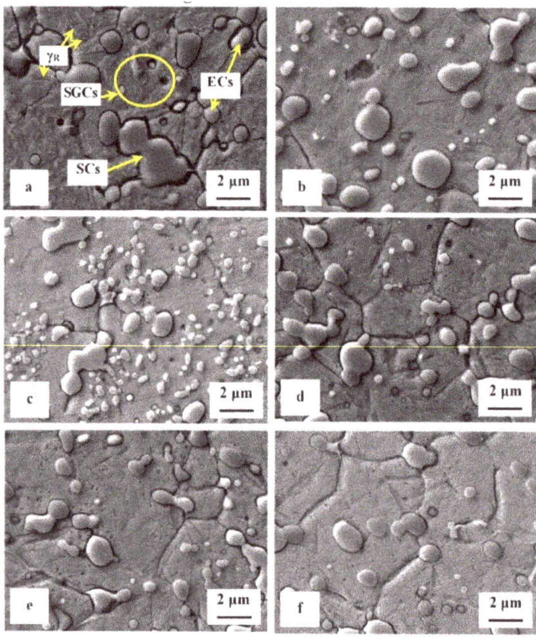

Figure 2. SEM micrographs showing the microstructure of no-tempered steel after conventional heat treatment (**a**), and after SZT at −75 °C for 4 h (**b**), 10 h (**c**), 17 h (**d**), 24 h (**e**), and for 48 h (**f**).

Table 2. Determined values of population density of small globular carbides for differently sub-zero treated and tempered specimens.

Duration of SZT [h]	Tempering Temperature [°C]				
	No	170	330	450	530
CHT	38.7	33.5	35.3	47.6	37.4
4	171.6	119.4	116.1	107.9	95.4
10	233.2	125.6	97.0	84.5	73.5
17	136.9	121.9	101.7	103.0	93.3
24	121.8	104.6	84.8	73.7	69.2
48	119.8	115.6	115.4	106.0	79.5

It is shown in Table 2 and in Figure 3 that tempering treatment always induces a reduction of population density of SGCs. Despite that their population density remains two- to three-fold higher than what is produced by conventional heat treatment. Finally, it should be mentioned that SZT does not modify the amounts and population densities of ECs and SCs in Vanadis 6 steel as reported recently [7,12]. The reason is that these carbides are stable up to much higher temperatures and thus neither the SZTs nor the tempering do modify their quantitative characteristics.

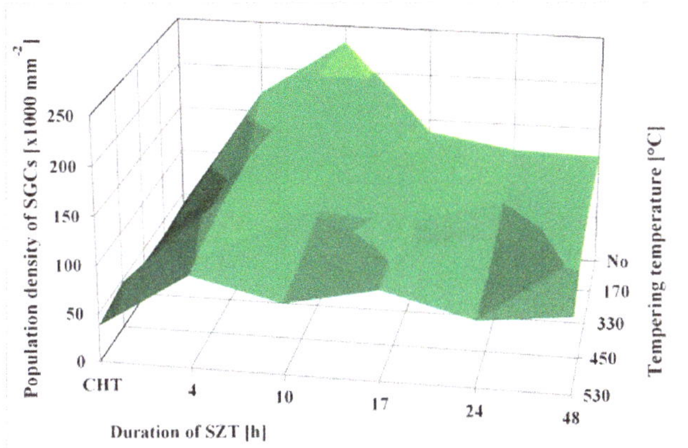

Figure 3. Population density of small globular carbides for differently sub-zero treated (at −75 °C) and tempered specimens.

A series of high-magnification SEM micrographs, Figure 4, depicts the microstructural alterations of sub-zero treated (at −75 °C) steel with increasing the tempering temperature. The as-sub-zero treated material microstructure contains the matrix formed by the martensite and retained austenite and eutectic, secondary, and small globular carbides, Figure 4a. Tempering treatment modifies the material microstructure as follows: The matrix manifests more pronounced sensitivity to the etching agent, Figure 4b–e. This is reflected by extensive roughening of the originally smooth metallographic surface, and thus by lowered distinctness of originally clearly visible matrix microstructural features (martensite laths, retained austenite, grain boundaries etc.) obtained by austenitizing, quenching, and sub-zero treatments, compare with Figure 4a. Mentioned changes can be ascribed to extensive precipitation of nano-sized carbides. These carbide particles become visible after tempering at 450 or 530 °C, as tiny elongated formations, Figure 4d,e. At the same time the retained austenite disappears from the microstructure because it was decomposed into either "secondary" martensite or bainite during cooling down from the tempering temperature. Last but not the least, it should be noticed the population density of both the ECs and the SCs are nearly unaffected by tempering while the population density of SGCs decreases with tempering treatment, compare Figure 4a with micrographs in Figure 4b–e, and see also Table 2. It should be noted that similar microstructural development was detected also for the steel after SZT at −75 °C for other (shorter or longer) durations.

The dependence of the retained austenite amount on the tempering temperature for specimens after sub-zero treatment at −75 °C for 17 h is in Table 3. The amount of γ_R was 7.6 ± 0.4 vol% in the prior-to-tempering state. Tempering at low temperatures reduces the γ_R amount only moderately. In contrast, tempering at higher temperatures results in either the significant reduction (450 °C) or in the almost complete removal of the retained austenite.

Table 3. Retained austenite amount (in vol%) in Vanadis 6 steel after quenching followed by sub-zero treatment at −75 °C for 17 h, no-tempered and tempered at different temperatures.

Tempering Temperature [°C]	No-Tempered	170	330	450	530
Retained austenite amount [vol%]	7.6 ± 0.4	5.5 ± 0.1	4.6 ± 0.2	2.5 ± 0.1	Not measurable

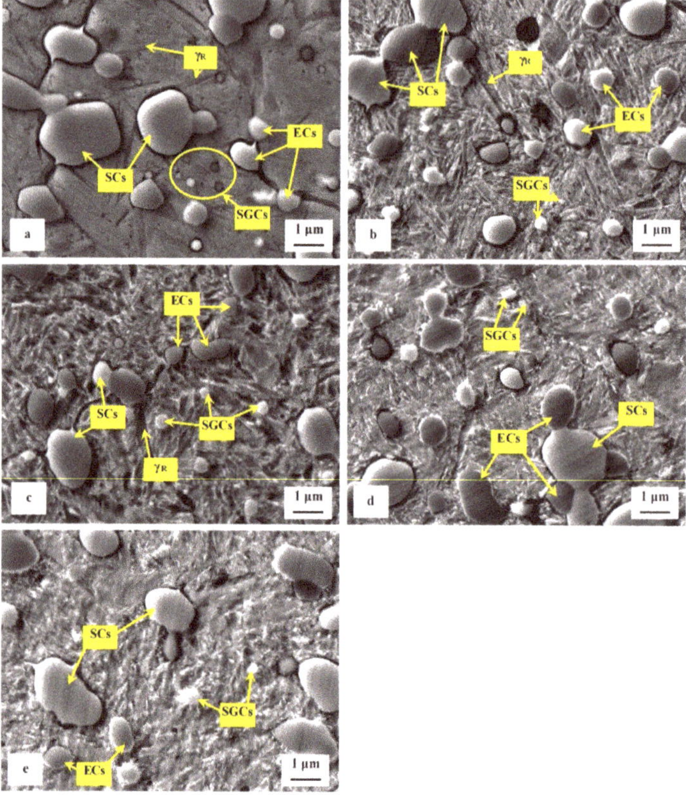

Figure 4. High-magnification SEM micrographs showing the microstructure of SZT steel (for 17 h) in the state after SZT (**a**), and after subsequent tempering at 170 °C (**b**), 330 °C (**c**), 450 °C (**d**), and 530 °C (**e**).

The γ_R amount in CHT specimens was around 20 vol% as reported recently [7,12]. The obtained results imply that the SZT at −75 °C reduces the γ_R amount to approximately one third as compared with the room temperature quenching. At the same time, however, it is worth to note that treatments at −140 or −196 °C act more effectively in reduction of retained austenite; the γ_R amounts were determined 4.3 and 2.1 vol% for steel that was SZT at −140 and −196 °C, respectively [7,12,26].

The prior-to-tempering hardness for conventionally (room temperature) quenched Vanadis 6 steel was 875 ± 16 HV10, Figure 5. The hardness values of the steel after sub-zero treatment at −75 °C for 4, 17, and 48 h were 908 ± 30, 930 ± 10, and 925 ± 6 HV10, respectively. The obtained results, thus, imply that the hardness of the Vanadis 6 steel is higher due to the sub-zero treatments at −75 °C, and that the hardness improvement reaches the maximum for the material treated for the duration of 17 h. On the other hand, the hardness improvement is less significant than what was obtained by the treatment at lower SZT temperatures, e.g. at −140 or −196 °C, where values exceeding 950 HV10 were obtained [12,22].

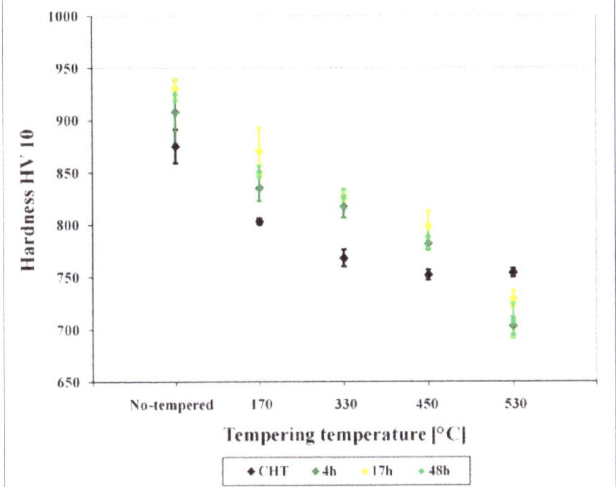

Figure 5. Hardness versus tempering temperature plots for conventionally heat-treated specimens and for specimens after sub-zero treatments at −75 °C for different durations.

The hardness of conventionally quenched steel was 803, 768, 752, and 754 HV 10 after tempering at temperatures of 170, 330, 450, and 530 °C, respectively. In other words, the hardness of conventionally heat-treated steel first decreases with increasing tempering temperature, and then it is preserved, almost constantly (at a level of 750 HV 10), when tempered at temperatures normally used for secondary hardening. In contrast, the steel after application of sub-zero treatments manifested higher hardness values, and this tendency was maintained up to the tempering temperature of 450 °C. The most significant hardness improvement was recorded for the steel treated for 17 h. For the specimens tempered at 530 °C, however, the hardness of sub-zero treated steel was lower than what was obtained by CHT.

These results are in line with those obtained by investigations of tempering response of the steel after SZT at −196 °C [12,20,21]. On the other hand, a strong variance between the tempering responses of the steel treated at −75 °C and the steel treated at −140 °C is evidenced. For the latter SZT temperature it has been reported recently that hardness of the material was improved significantly within the whole range of tempering regimes used [22].

The flexural strength values obtained by three-point bend tests of samples having the microstructures according to Figure 4 are shown in Figure 6. It can be seen that the mean values of the flexural strength range between 3300 and 3700 MPa. The ranges of statistical uncertainty (at a probability level of 5%) overlap noticeably, suggesting that the tempering has only marginal effect on the flexural strength of SZT steel.

The work of fracture values of the steel tempered at 170, 330 and 450 °C lie within the range of statistical uncertainty, Figure 6, suggesting that low-temperature tempering does not influence this characteristic significantly. On the other hand, the work of fracture increases when the steel is tempered at temperature 530 °C and above. This can be associated with hardness decrease (as indicated in Figure 6), and thus with more extensive (compared to what occur during testing of specimens tempered at lower temperatures) plastic deformation of the matrix.

It is also well visible that the W_{of} values follow closely the values of flexural strength, but only up to the tempering temperature of 530 °C (peak of secondary hardness). For the specimens tempered at 600 °C, however, this relationship is no more valid. At this place it should be noted that the Vanadis 6 steel belongs to the group of brittle steels when tempered at temperatures up to the secondary hardness peak; in these cases, the obtained values of flexural strength are an indirect measure of its toughness. On the other hand, the flexural strength loses the characteristic of toughness when the hardness of the steel decreases. This is the case of "overtempering" of the steel, i.e., when the steel is tempered at temperatures beyond the secondary hardness peak, for instance at 600 °C.

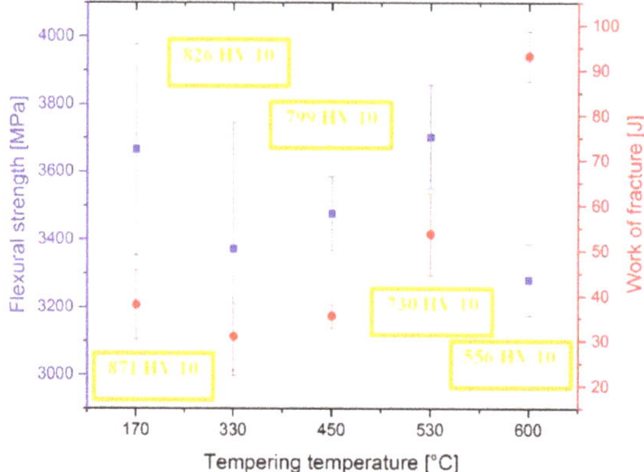

Figure 6. Flexural strength and work of fracture obtained by flexural three-point bend test for steel with application of SZT at −75 °C for 17 h, and subsequent tempering.

Figure 7 is a compilation of SEM micrographs showing fracture surfaces of flexural strength specimens that were sub-zero treated at −75 °C for 17 h and subsequently tempered at different temperatures. With respect to the surface morphology, the fracture surfaces can be divided into two groups. The fracture surfaces of the first group of specimens appear relatively flat, with only very limited roughness caused by strong difference between the fracture behavior of the matrix and carbides, Figure 7a–c. In contrast, the fracture surfaces (second group) of the specimens tempered at 530 and 600 °C (the latter one in particular) manifest much more pronounced indications of plastic deformation of the matrix, Figure 7d,e. Here it should be noted that the steel hardness decreased (and the plasticity expectedly increased) with increasing the tempering temperature, Figure 6. This correlates well with the above-mentioned morphology of fracture surfaces. In other words, the fracture surfaces appear relatively flat and shiny when the steel has higher hardness, but their topography increases with hardness decrease (i.e., with increasing the tempering temperature).

Detailed SEM fractographs in Figure 8 clearly delineate the differences between the fracture surfaces obtained by testing of specimens tempered at 170 and 600 °C. A relatively flat morphology of the fracture surface with only very limited plastic deformation of the matrix is visible in Figure 8a. The presence of micro-plastic deformation is only visible at the sites where decohesion at the matrix/carbide interface took place during the crack propagation. The sites with local micro-plastic deformation are located mainly in close vicinity to smaller carbides, which act as decohesion sites at the above-mentioned interfaces. These carbides are denoted as decohesive carbides (DCs). Other carbide particles (the coarser ones, in most cases) are cleaved, and they are denoted as "cleaved carbides, CCs". In contrast to the fracture surface shown in Figure 8a the fracture of the specimen tempered at 600 °C manifests much more pronounced plastic deformation, Figure 8b. However, the role of particular carbides in the crack propagation is maintained and is almost the same as shown in Figure 8a.

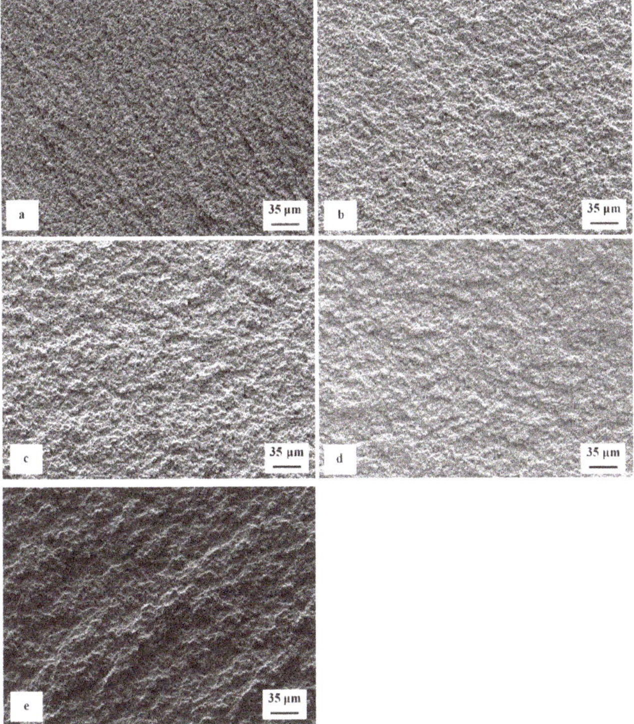

Figure 7. Representative SEM micrographs of the fracture surfaces of sub-zero treated specimens after flexural strength testing, (**a**) tempered at 170 °C, (**b**) tempered at 330 °C, (**c**) tempered at 450 °C, (**d**) tempered at 530 °C, and (**e**) tempered at 600 °C.

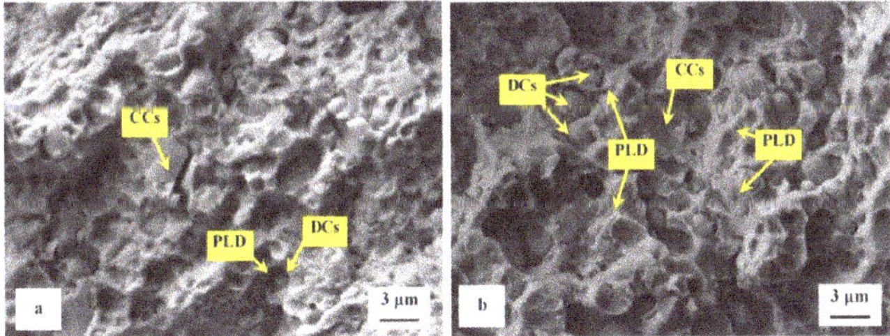

Figure 8. Detailed SEM micrographs of the fracture surfaces of sub-zero treated specimens after flexural strength testing, (**a**) tempered at 170 °C and (**b**) tempered at 600 °C.

The obtained flexural strength values for differently sub-zero treated and tempered specimens are summarized in Figure 9. It is shown that the treatment at −75 °C gave rather higher flexural strength than the conventional heat treatment. This is somewhat surprising at first glance, since one would expect a decrease in flexural strength rather than slight increase due to the application of SZT.

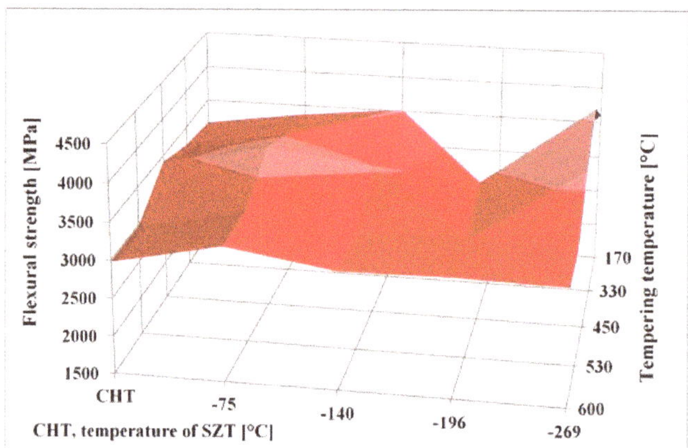

Figure 9. Comparison of the obtained flexural strength values for differently sub-zero treated and tempered specimens made of the Vanadis 6 steel.

To explain this, it should be noted that Vanadis 6 steel contains 20.2, 7.6 4.8, 2.1, and 6.3 vol % of retained austenite after CHT, SZT at −75 °C, SZT at −140 °C, SZT at −196 °C, and SZT at −269 °C, respectively (all the SZTs were carried out for 17 h duration) [12,26]. Retained austenite is considered a soft microstructural feature, and one can expect its beneficial effect on toughness (and flexural strength). This may, for instance, partly explain the better flexural strength of the conventionally heat-treated material, as well as the material SZT at either −140 or −269 °C, than was obtained by SZT at −196 °C. However, this does not bring an answer to the improvement of flexural strength by SZT at −75 °C, compared to CHT steel. As mentioned above, sub-zero treatment at −75 °C produces a much higher population density of small globular carbides than CHT, which in turn leads to the formation of an increased number of matrix/carbide interfaces. The population density of small globular carbides decreases with tempering, despite remaining much higher than what can be obtained by CHT. In the steel samples with a higher population density of carbides, the crack propagation is more probably associated with micro-plastic deformation of a matrix (as Figure 8 clearly demonstrates); hence, an enhanced carbide count is an important factor responsible for improvements of flexural strength.

Finally, few words should be written to the fact that the treatments at either −140 or −269 °C lead to nearly equal or slightly better flexural strength than what is obtained in the present study. It has been reported recently that the treatment at −140 °C for 17 h results in the retained austenite amount of around 4.3 vol% [22]. The latest measurements fixed the γ_R amount to 6.3 vol% in the Vanadis 6 steel treated in liquid helium. This is two times (for −140 °C SZT) or three times (for −269 °C) more than what was obtained by SZT at −196 °C, but only of approximately 60% or 85% in comparison with the values reported here for SZT at −75 °C. One can thus expect rather opposite tendency with respect to the changes of flexural strength; however, the treatment at −140 °C (for instance) produces the greatest population density of SGCs, which more than fully compensates the toughness loss resulting from the reduction of retained austenite.

The relation between fracture toughness and hardness, as a function of tempering temperature, is in Figure 10. The of fracture toughness values were 13.76 ± 0.61, 17.60 ± 0.75, 16.40 ± 0.32, 15.35 ± 0.32, and 17.04 ± 0.15 MPa × m$^{1/2}$ for specimens prior-to-tempering, tempered at 170, 330, 450, and 530 °C, respectively. The lowest K_{IC} values were determined for prior-to-tempering specimens, which had the highest hardness. This is logical because the hardness is the key factor that influences the fracture toughness, and it is generally accepted that metals with high hardness usually manifest low fracture toughness level, and vice versa [20,33]. The values of K_{IC} and hardness that were obtained by testing of tempered specimens do not obey this rule, however. It is for instance shown that the fracture toughness is high for the steel tempered at 170 °C despite its high hardness. Further, the K_{IC} decreases along with the hardness when the tempering temperature is increased, to either 330 or 450 °C. An increase in K_{IC} was recorded only for the steel tempered at 530 °C, but at lower a hardness value.

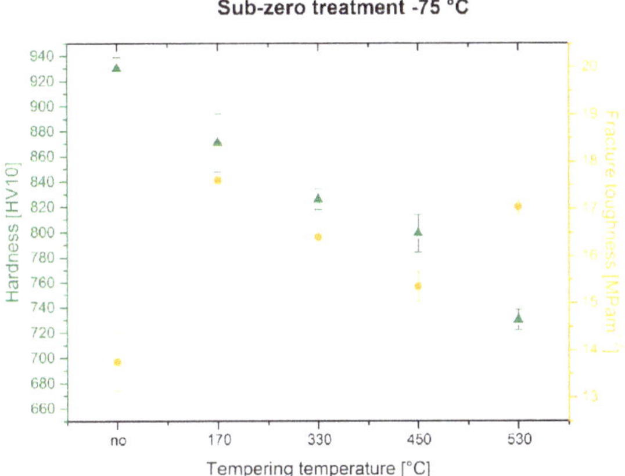

Figure 10. Fracture toughness versus hardness of the Vanadis 6 steel after SZT at −75 °C for 17 h, and subsequent tempering.

The analysis of fracture surfaces provides a more comprehensive insight into the variations of fracture toughness with tempering. Figure 11 is a compilation of representative SEM micrographs of the fractured K_{IC} specimens that were SZT at −75 °C and no-tempered or tempered at different temperatures. The fracture surfaces of all the specimens manifest symptoms that are typical for hard and brittle steels; they appear flat, shiny, and relatively smooth. However, more thorough investigation reveals differences on the topography of fractured surfaces. The surface of no-tempered specimen (Figure 11a) manifests much finer topography compared with the fractured surfaces of other specimens (Figure 11b–e). This can be attributed to the differences in fracture toughness-no-tempered specimens have the lowest K_{IC} values, and tempering leads to moderate increase in this characteristic.

Detailed SEM micrographs in Figure 12 were acquired from the specimen that was not tempered after sub-zero treatment. The image in Figure 12a assists in seeing that the fracture surface manifests typical morphology for hard steel. It contains a great number of micro-voids and holes, which correspond to extraction of SGCs (in particular) from the fracture surface during the crack propagation. The formation of micro-voids is associated with local plastic deformation of the matrix. However, the capability of the matrix to be deformed plastically is very limited, Figure 12b. There is a great number of sites with cleavage fracture mechanism apparent in the matrix as Figure 12c illustrates. The micrograph in Figure 12b also depicts the difference in behavior of carbides during the crack propagation. Some carbide particles (mainly the coarsest ones) were cleaved while others assisted the decohesion mechanism of the crack propagation. This is similar to what was discovered for the flexural strength specimens, Figure 8.

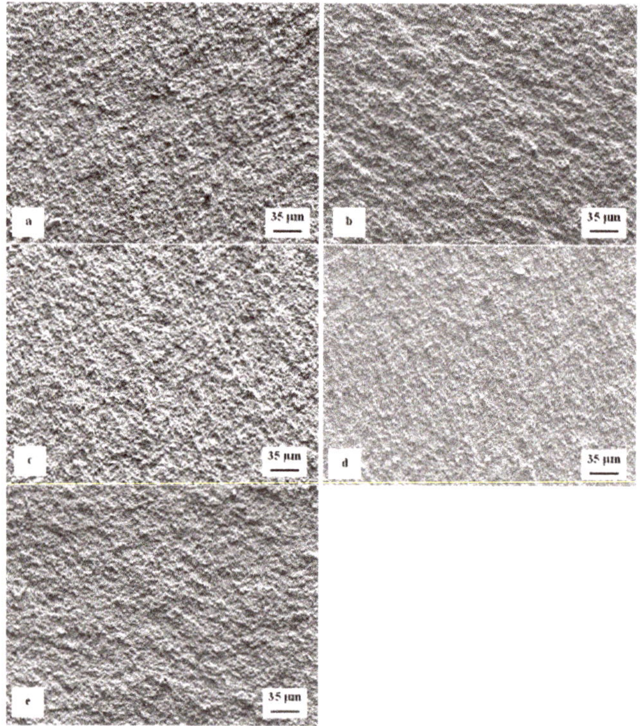

Figure 11. Representative SEM micrographs of the fracture surfaces of sub-zero treated specimens after fracture toughness testing, (**a**) no-tempered, (**b**) tempered at 170 °C, (**c**) tempered at 330 °C, (**d**) tempered at 450 °C, and (**e**) tempered at 530 °C.

The role of particular carbides in the fracture propagation can be assessed with the help of recent investigations. It has been published recently that the coarsest particles belong mostly to the group of secondary carbides and that their nature is hexagonal M_7C_3 [7,11,12,22]. The finer particles are either eutectic carbides (MC-carbides with cubic crystallographic structure) or small globular carbides (cementite M_3C) [7]. Casellas et al. clearly demonstrated that lower symmetry of the hexagonal crystalline lattice (as compared with cubic MC phase) results in low fracture toughness of M_7C_3, and correspondingly in large scatter of its values (0.5–4.5 MPa $\times$ m$^{1/2}$) [34]. Moreover, Fukaura et al. [35] proved that the size of the carbides plays an important role in their fracture propagation manner, namely that coarser particles are much more amenable to the cleavage than the finer carbides. In Vanadis 6 steel, the mean spherical diameter of the M_7C_3 particles is of around 2.5–2.8 µm while the MC carbides have size within the range 1.6–1.9 µm [26]. The size of small globular carbides, SGCs, is even much smaller, below 0.5 µm. These facts assist to delineate the role of different carbides in fracture propagation as Figure 12 illustrates. Larger size and crystallographic anisotropy of M_7C_3 make this carbide more brittle than the MC (or M_3C), despite that the hardness of M_7C_3 is lower than that of MC [34]. This is why the most part of cleaved carbides are the SCs (M_7C_3) while dominant number of ECs (and almost all the SGCs) is unaffected by the crack propagation, thus, assisting decohesion at the matrix/carbide interfaces.

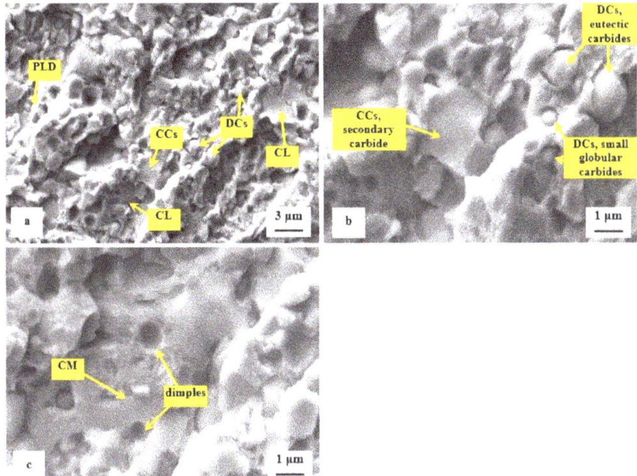

Figure 12. SEM images showing details of fracture propagation in sub-zero treated and no-tempered fracture toughness specimen, (**a**) detailed image showing the crack propagation in the matrix and the three main carbides including their role in fracture propagation, (CCs—cleaved carbides, DCs—decohesive carbides, PLD—plastic deformation in the matrix, CL—cleaved matrix), (**b**) high-magnification image showing cleaved M_7C_3 (CCs, secondary carbide), and decohesive particles that belong to eutectic carbides (DCs, eutectic carbides) and small globular carbides, and (**c**) high-magnification image showing cleaved matrix region (CM) and small dimples resulting from the extraction of small globular carbides.

The second series of detailed SEM micrographs, Figure 13, depicts the details of fracture propagation in the specimen that was sub-zero treated and subsequently tempered at 530 °C. It is worth noting that the fractured surfaces of specimens tempered at other temperatures did not differ significantly from that in Figure 13. Compared to the fracture surface of prior-to-tempering specimen, Figure 12, the fracture surface of the tempered one contains much greater area fraction of micro-plastically deformed matrix, Figure 13a. Further, there are cleaved carbides (CCs) and decohesive carbides present.

High-resolution SEM micrograph in Figure 13b shows details of cleaved secondary carbides (CCs) and decohesive eutectic carbide particles (ECs), suggesting that the role of particular carbides in the fracture propagation is very similar to the case of the prior-to tempering specimen, see Figure 12. In other words, the difference in fracture toughness (13.76 ± 0.61 MPa × m$^{1/2}$ for the prior-to tempering steel versus 17.04 ± 0.15 MPa × m$^{1/2}$ for the steel tempered at 530 °C) does not play a significant role in the fracture propagation mode of carbide particles. In contrast, mentioned difference in fracture toughness values is reflected in the matrix behavior. As Figure 13c illustrates, the fracture propagation on the matrix is associated with micro-plastic deformation, and the cleavage takes only a minor role.

The differences in the fracture toughness, and correspondingly in the morphology of fracture surfaces of differently tempered specimens can be explained considering the material microstructure. Table 3 shows that the examined steel contains 7.6 ± 0.4 vol% of retained austenite in the prior-to-tempering state. The retained austenite is relatively soft, thus prone to plastic deformation. For instance, Putatunda [33] reported that the retained austenite has beneficial effect on the fracture toughness of high-carbon steels. However, the Vanadis 6 steel also contains high amount of hard and brittle "pre-aged" martensite in the prior-to-tempering state (it is worth noting that aging of the martensite in sub-zero treated high-carbon steels was many times experimentally proved, e.g. [14,15,26,36]), which makes the steel brittle. Berns and Broeckmann [1], Das et al. [19], and Ptačinová et al. [21] found out that the crack has a strong tendency to follow the interfaces between matrix and carbides when propagates during testing of ledeburitic steels. As a result, a micro-plastic deformation of the matrix occurs, which slightly improves the fracture toughness. Nevertheless, beneficial effects of retained austenite and increased population density of carbides cannot compensate the high brittleness of pre-aged martensite, hence, the material manifests very low fracture toughness as a consequence. Finally, it should be underlined that this finding is in line with the obtained results on the same steel, but processed at different SZT temperatures as Figure 14 illustrates. Finally, it should be underlined that this finding is in line with the obtained results by the K_{IC} testing of the same steel that was processed at other SZT temperatures as Figure 14 illustrates.

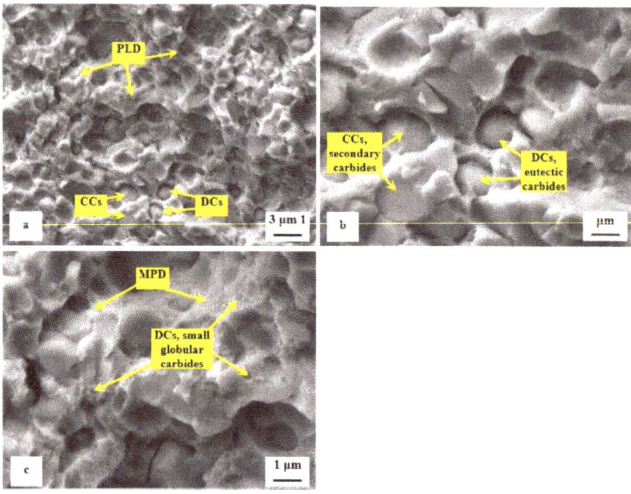

Figure 13. SEM images showing details of fracture propagation in sub-zero treated and subsequently tempered (at 530 °C) fracture toughness specimen, (**a**) detailed image showing the crack propagation in the matrix and the three main carbides including their role in fracture propagation, (CCs—cleaved carbides, DCs—decohesive carbides, PLD—plastic deformation in the matrix), (**b**) high-magnification image showing cleaved M_7C_3 (CCs, secondary carbide), and decohesive particles that belong to eutectic carbides (DCs, eutectic carbides), and (**c**) high-magnification image showing micro-plastic deformation of the matrix (MPD) and small dimples resulting from the extraction of small globular carbides.

Tempering reduces the material hardness, Figure 5, and one can thus expect increase of fracture toughness. Increased fracture toughness was really recorded for sub-zero treated and subsequently tempered steel specimens, Figure 10. The clarification of this issue is relatively complex: On one hand the amount of soft retained austenite is either moderately reduced (for the steel tempered at 170 or 330 °C) or this phase is almost completely removed (after application of higher tempering temperatures), Table 3. Also, tempering reduces the population density of small globular carbides, Table 2, Figure 3. On the other hand, the martensite undergoes significant softening, which makes it more amenable to deform plastically. The resulting fracture toughness values are then a result of competition between mentioned three phenomena. The increase of K_{IC} values can be mainly ascribed to the martensite softening as this phase is the major one in the material. Increase in fracture toughness is reflected in more pronounced topography of fractured surfaces (and in more visible dimples on fracture surfaces at the same time), compare Figure 11a with Figure 11b–e, or Figure 12 with Figure 13.

Very interesting comparison of the current results with those obtained by testing of CHT steel and the steel that was subjected to SZT at −140, −196, or −269 °C is provided in Figure 14. It is shown that the SZT at −140 °C gave the best results that are fully comparable with conventionally heat-treated steel, but at significantly increased hardness as reported in [22]. Treatments at either the temperatures of boiling nitrogen or helium resulted rather in lower fracture toughness values.

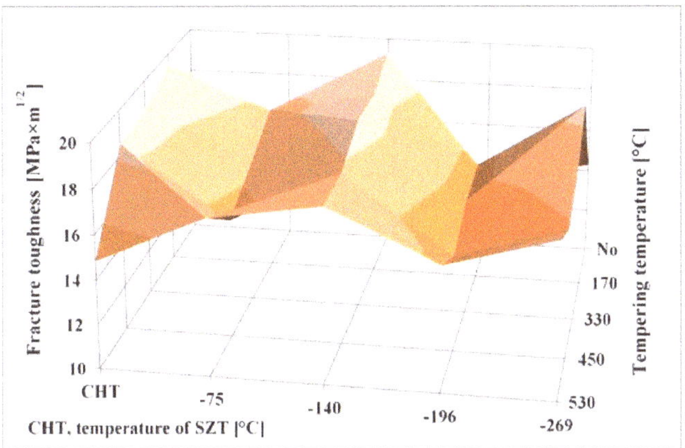

Figure 14. Comparison of the obtained fracture toughness values for differently sub-zero treated and tempered specimens made of the Vanadis 6 steel.

A reliable explanation of the variations in fracture toughness is very complex issue. In brief, the resulting fracture toughness value of ledeburitic tool steels is always a result of competition between three effects: i) retained austenite amount, which undoubtedly acts in favor of higher K_{IC}, ii) state of the martensite (K_{IC} decrease), and iii) population density of small globular carbides (increase of K_{IC}) [22]. CHT steel contains of around 20 vol% of the retained austenite in the prior-to-tempering state, and the retained austenite amount is nearly constant up to the tempering temperature of 500 °C [12]. Even though the population density of SGCs is very low in CHT steel (as compared with the steel after SZTs) it is more than satisfactorily compensated by the retained austenite amount. Application of SZTs reduces the retained austenite amount considerably (with the minimum value for SZT at −196 °C) and increases the population density of SGCs (with the maximum for the SZT at −140 °C). The results in Figure 14 imply that the reduction of retained austenite is almost fully compensated by much higher population density of SGCs for the steel after SZT at −140 °C while other SZTs do not lead to high enough population density of carbides needed to compensate the impact of reduced retained austenite amount on the resulting K_{IC} value.

Finally, it should be noted that these considerations do not take into account the state of the martensite. In the state after CHT (and prior-to-tempering) the martensite was found to be "pre-aged", but it did not contain any carbide precipitates [7]. In contrast, the martensite after SZTs contained nano-sized, coherent transient carbides ([12,26]. However, it is hard or practically impossible to provide an exact assessment of the difference between impacts of these two martensite states on the fracture toughness of the steel since it also contains the retained austenite and several carbide types. The situation is very similar in the case of tempered steel Vanadis 6 because tempering leads to acceleration of precipitation rate at low-tempering temperatures but rather to delayed precipitation at high temperatures (around 500 °C) [11,12,26]. But, an exact quantification of precipitates is almost impossible.

4. Conclusions

The impact of sub-zero treatment at the temperature of −75 °C and subsequent tempering on the microstructure, hardness, flexural strength and fracture toughness was investigated. The main obtained results can be summarized as follows:

- The population density of small globular carbides is increased by application of SZT at −75 °C, by the factor from the range 2.5 and seven.

- Retained austenite amount is reduced by this kind of treatment, to an approximately one third as compared with conventional room temperature quenching.
- Third bullet
- The bulk hardness of the steel is increased by SZT at −75 °C, by 30–60 HV10. Improved steel hardness is maintained up to the tempering temperature of 450 °C while tempering at 530 °C leads to more significant hardness reduction than what is obtained by conventional heat treatment.
- The flexural strength of SZT steel ranges between 3300 and 3700 MPa. The level of tempering temperature has only little impact in the flexural strength.
- The fracture toughness of sub-zero treated no-tempered steel is very low. However, it increases with the tempering application, and reaches relatively high values of around 17 MPa × m1/2 after tempering to the secondary hardness peak.
- In summary, the application of SZT at −75 °C may bring some benefits into heat treatment of tool steels. However, the obtained microstructures and values of mechanical properties are lower as compared, for instance, with those obtained by treatment at −140 °C.

Author Contributions: Funding acquisition, P.J.; Microstructural investigation, L.R.-T., P.J.; X-ray diffraction, M.K.; Dilatometry. J.K.; Flexural strength, fracture toughness testing, I.D.; Supervision, P.J.; Writing—original draft, review, editing, P.J., I.D.

Funding: The authors acknowledge that the paper is a result of experiments realized within the project VEGA 1/0264/17. In addition, this publication is the result of the project implementation "Centre for Development and Application of Advanced Diagnostic Methods in Processing of Metallic and Non-Metallic Materials – APRODIMET", ITMS: 26220120014, supported by the Research & Development Operational Programme funded by the ERDF.

Conflicts of Interest: The authors declare no conflict of interest.

References

1. Berns, H.; Broeckmann, C. Fracture of Hot Formed Ledeburitic Chromium Steels. *Eng. Fract. Mech.* **1997**, *58*, 311–325. [CrossRef]
2. Das, D.; Ray, K.K. Structure-property correlation of cub-zero treated AISI D2 steel. *Mater. Sci. Eng. A* **2012**, *541*, 45–60. [CrossRef]
3. Amini, K.; Akhbarizadeh, A.; Javadpour, S. Investigating the effect of the quench environment on the final microstructure and wear behaviour of 1.2080 tool steel after deep cryogenic heat treatment. *Mater. Des.* **2013**, *45*, 316–322. [CrossRef]
4. Das, D.; Ray, K.K.; Dutta, A.K. Influence of temperature of sub-zero treatments on the wear behaviour of die steel. *Wear* **2009**, *267*, 1361–1370. [CrossRef]
5. Surberg, C.H.; Stratton, P.F.; Lingenhoele, K. The effect of some heat treatment parameters on the dimensional stability of AISI D2. *Cryogenics* **2008**, *48*, 42–47. [CrossRef]
6. Akhbarizadeh, A.; Javadpour, S.; Amini, K.; Yaghtin, A.H. Investigating the effect of ball milling during the deep cryogenic heat treatment of the 1.2080 tool steel. *Vacuum* **2013**, *90*, 70–74. [CrossRef]
7. Jurči, P.; Dománková, M.; Čaplovič, Ľ.; Ptačinová, J.; Sobotová, J.; Salabová, P.; Prikner, P.; Šuštaršič, B.; Jenko, D. Microstructure and hardness of sub-zero treated and no tempered P/M Vanadis 6 ledeburitic tool steel. *Vacuum* **2015**, *111*, 92–101. [CrossRef]
8. Amini, K.; Akhbarizadeh, A.; Javadpour, S. Investigating the effect of holding duration on the microstructure of 1.2080 tool steel during the deep cryogenic treatment. *Vacuum* **2012**, *86*, 1534–1540. [CrossRef]
9. Tyshchenko, A.I.; Theisen, W.; Oppenkowski, A.; Siebert, S.; Razumov, O.N.; Skoblik, A.P.; Sirosh, V.A.; Petrov, J.N.; Gavriljuk, V.G. Low-temperature martensitic transformation and deep cryogenic treatment of a tool steel. *Mater. Sci. Eng. A* **2010**, *527*, 7027–7039. [CrossRef]
10. Villa, M.; Pantleon, K.; Somers, M.A.J. Evolution of compressive strains in retained austenite during sub-zero Celsius martensite formation and tempering. *Acta Mater.* **2014**, *65*, 383–392. [CrossRef]
11. Jurči, P. Sub-Zero Treatment of Cold Work Tool Steels–Metallurgical Background and the Effect on Microstructure and Properties. *HTM J. Heat Treat. Mater.* **2017**, *72*, 62–68. [CrossRef]

12. Jurči, P.; Dománková, M.; Hudáková, M.; Ptačinová, J.; Pašák, M.; Palček, P. Characterization of microstructure and tempering response of conventionally quenched, short- and long-time sub-zero treated PM Vanadis 6 ledeburitic tool steel. *Mater. Charact.* **2017**, *134*, 398–415. [CrossRef]

13. Stratton, P.F. Optimizing nano-carbide precipitation in tool steels. *Mater. Sci. Eng. A* **2007**, *449–451*, 809–812. [CrossRef]

14. Van Genderen, M.J.; Boettger, A.; Cernik, R.J.; Mittemeijer, E.J. Early Stages of Decomposition in Iron-Carbon and Iron-Nitrogen Martensites: Diffraction Analysis Using Synchrotron Radiation. *Metall. Trans. A* **1993**, *24*, 1965–1973. [CrossRef]

15. Preciado, M.; Pellizzari, M. Influence of deep cryogenic treatment on the thermal decomposition of Fe-C martensite. *J. Mater. Sci.* **2014**, *49*, 8183–8191. [CrossRef]

16. Collins, D.N.; Dormer, J. Deep Cryogenic Treatment of a D2 Cold-work Tool Steel. *Heat Treat. Met.* **1997**, *3*, 71–74.

17. Wierszyllowski, I. The Influence of Post-quenching Deep Cryogenic Treatment on Tempering Processes and Properties of D2 Tool Steel. Studies of Structure, XRD, Dilatometry, Hardness and Fracture Toughness. *Defect Diffus. Forum* **2006**, *258–260*, 415–420.

18. Rhyim, Y.M.; Han, S.H.; Na, Y.S.; Lee, J.H. Effect of Deep Cryogenic Treatment on Carbide Precipitation and Mechanical Properties of Tool Steel. *Solid State Phenom.* **2006**, *118*, 9–14. [CrossRef]

19. Das, D.; Sarkar, R.; Dutta, A.K.; Ray, K.K. Influence of sub-zero treatments on fracture toughness of AISI D2 steel. *Mater. Sci. Eng. A* **2010**, *528*, 589–603. [CrossRef]

20. Sobotová, J.; Jurči, P.; Dlouhý, I. The effect of sub-zero treatment on microstructure, fracture toughness, and wear resistance of Vanadis 6 tool steel. *Mater. Sci. Eng. A* **2016**, *652*, 192–204. [CrossRef]

21. Ptačinová, J.; Sedlická, V.; Hudáková, M.; Dlouhý, I.; Jurči, P. Microstructure Toughness relationships in sub-zero treated and tempered Vanadis 6 steel compared to conventional treatmen. *Mater. Sci. Eng. A* **2017**, *702*, 241–258. [CrossRef]

22. Jurči, P.; Ďurica, J.; Dlouhý, I.; Horník, J.; Planieta, R.; Kralovič, D. Application of −140 °C Sub-Zero Treatment For Cr-V Ledeburitic Steel Service Performance Improvement. *Metall. Mater. Trans. A* **2019**, *50*, 2413–2434. [CrossRef]

23. Kumar, S.; Nahraj, M.; Bongale, A.; Khedkar, N.K. Effect of deep cryogenic treatment on the mechanical properties of AISI D3 tool steel. *Int. J. Mater. Eng. Innov.* **2019**, *10*, 98–113. [CrossRef]

24. Reitz, W.; Pendray, J. Cryoprocessing of Materials: A Review of Current Status. *Mater. Manuf. Process.* **2001**, *16*, 829–840. [CrossRef]

25. Gavriljuk, V.G.; Theisen, W.; Sirosh, V.V.; Polshin, E.V.; Kortmann, A.; Mogilny, G.S.; Petrov, Y.N.; Tarusin, Y.V. Low-temperature martensitic transformation in tool steels in relation to their deep cryogenic treatment. *Acta Mater.* **2013**, *61*, 1705–1715. [CrossRef]

26. Ďurica, J.; Ptačinová, J.; Dománková, M.; Čaplovič, L.; Čaplovičová, M.; Hrušovská, L.; Malovcová, V.; Jurči, P. Changes in microstructure of ledeburitic tool steel due to vacuum austenitizing and quenching, sub-zero treatments at −140°C and tempering, Vacuum 2019. **2019**. [CrossRef]

27. Zurecki, Z. Cryogenic Quenching of Steel Revisited. In Proceedings of the 23rd Heat Treating Society Conference, Pittsburgh, PA, USA, 26–28 September 2005; pp. 106–114.

28. ASTM E975-13. *Standard Practice for X-Ray Determination of Retained Austenite in Steel with Near Random Crystallographic Orientation*; ASTM Book of Standards: West Conshohocken, PA, USA, 2004; Volume 3.01.

29. CSN EN ISO 6507-1. *Metallic Materials - Vickers Hardness Test - Part 1: Test Method*; Office for Technical Standardization, Metrology and State Testing: Prague, Czech Republic, 2018.

30. Jurči, P.; Dlouhý, I. Fracture Behaviour of P/M Cr-V Ledeburitic Steel with Different Surface Roughness. *Mater. Eng.* **2011**, *18*, 36–43.

31. CSN EN ISO. *Hardmetals – Determination of Transverse Rupture Strength*; Office for Technical Standardization, Metrology and State Testing: Prague, Czech Republic, 2009.

32. ISO 12135. Metallic Materials – Determination of Plane Strain Fracture Toughness. 2010. Available online: https://www.iso.org/standard/60891.html (accessed on 1 November 2016).

33. Putatunda, S.K. Fracture toughness of a high carbon and high silicon steel. *Mater. Sci. Eng. A* **2001**, *297*, 31–43. [CrossRef]

34. Casellas, D.; Caro, J.; Molas, S.; Prado, J.M.; Valls, I. Fracture toughness of carbides in tool steels evaluated by nanoindentation. *Acta Mater.* **2007**, *55*, 4277–4286. [CrossRef]

35. Fukaura, K.; Yokoyama, Y.; Yokoi, D.; Tsuji, N.; Ono, K. Fatigue of cold-work tool steels: Effect of heat treatment and carbide morphology on fatigue crack formation, life, and fracture surface observations. *Metall. Mater. Trans. A* **2004**, *35*, 1289–1300. [CrossRef]
36. Cheng, L.; Brakman, C.M.; Korevaar, B.M.; Mittemeijer, E.J. The tempering of iron-carbon martensite; dilatometric and calorimetric analysis. *Metall. Trans. A* **1988**, *19*, 2415–2426. [CrossRef]

Article

Study of the Catalytic Strengthening of a Vacuum Carburized Layer on Alloy Steel by Rare Earth Pre-Implantation

Guolu Li [1], Caiyun Li [1,2], Zhiguo Xing [2,*], Haidou Wang [2,*], Yanfei Huang [2], Weiling Guo [2] and Haipeng Liu [3]

[1] School of Materials Science and Engineering, Hebei University of Technology, Tianjin 300130, China; liguolu@hebut.edu.cn (G.L.); licaiyun0301@163.com (C.L.)

[2] National Key Laboratory for Remanufacturing, Academy of Army Armored Forces, Beijing 100072, China; huangyanfei123@126.com (Y.H.); guoweiling_426@163.com (W.G.)

[3] Process Institute of Inner Mongolia First Machinery Group Co, Ltd, Baotou Inner Mongolia 014030, China; sheb008@163.com

* Correspondence: xingzg2011@163.com (Z.X.); wanghaidou@aliyun.com (H.W.); Tel.: +86-010-6671-8540 (Z.X.); +86-010-6671-8541 (H.W.)

Received: 3 September 2019; Accepted: 15 October 2019; Published: 18 October 2019

Abstract: Conventional carburizing has disadvantages, such as high energy consumption, large deformation of parts, and an imperfect structure of the carburizing layer. Hence, a rare earth ion pre-implantation method was used to catalyze and strengthen the carburized layer of 20Cr2Ni4A alloy steel. In this study, X-ray diffraction (XRD), X-ray photoelectron spectroscopy (XPS), optical microscopy (OM), scanning electron microscopy (SEM), energy dispersive microanalysis (EDS), transmission electron microscopy (TEM), and Rockwell/Vickers hardness testing were used to analyze the microstructure, phase composition, retained austenite content, hardness, carburized layer thickness, and carbon diffusion. The results showed that lanthanum and yttrium ions implanted into the 20Cr2Ni4A steel formed solid solutions of rare earth ions and a large number of dislocations, which improved the diffusion coefficient of carbon elements on the carburized surface and the uniformity of the carbon distribution. Simultaneously, rare earth ion implantation improved the structure and hardness of the vacuum carburized layer. Compared to the lanthanum ion implantation, yttrium ion implantation caused the structure of the carburized layer to be finer, and the carbon diffusion coefficient increased by 1.17 times; in addition, the surface hardness of the carburized layer was 61.8 HRC.

Keywords: 20Cr2Ni4A; vacuum carburizing; ion implantation; rare earths; catalysis; carbon diffusion; microstructure

1. Introduction

Since 20Cr2Ni4A alloy steel has excellent surface properties after carburization, it has been widely used for gears and bearings in heavy-duty equipment that is frequently exposed to harsh conditions, such as heavy loads, elevated working temperatures, high frequency vibrations, and high rotational speeds [1,2]. Hence, failures such as wear, pitting, and contact fatigue damage, are more likely to occur on the surface of gears and bearings [3–5]. Vacuum carburization, as a surface strengthening method, plays an important role in the process of obtaining excellent comprehensive mechanical properties of 20Cr2Ni4A alloy steel. Vacuum carburization is a rapid process in which acetylene is injected into a high-temperature furnace in pulsed mode under low pressure and vacuum [6–8]. However, vacuum carburizing still has its disadvantages, such as a high carburizing temperature, a long carburizing cycle, an uneven structure, and coarse carbide particles. Studies showed that the introduction of rare earth (RE) ions in the carburizing heat treatment of steel or other parts reduces the treatment temperature and increases the carbon activity and the carbon diffusion coefficient during the carburization process; it can also improve the structure of the carburized layer [9,10]. Yan et al. [11,12] studied the effect of an RE catalyst on the gas carburizing kinetics of steel and found that compared to the conventional gas carburizing process, the incorporation of an RE led to an obvious increase in layer depth and carbon concentration profile, under the same carburizing conditions. The diffusion coefficient and transfer coefficient increased by 50% and 117%, respectively. Dong et al. [13] used RE ion implantation to assist vacuum carburizing of 12Cr2Ni4A steels. The results showed that the addition of RE elements obviously improved the hardness of the carburized layer and decreased the hardness gradient. Moreover, the structure of the carburized layer was compact, and the carbides were fine and dispersed.

Ion implantation is a modern material surface modification technology that has become increasingly common for changing the characteristics and properties of metal materials [14]. It can selectively improve the wear resistance, corrosion resistance, oxidation resistance, and fatigue resistance of the material surface by implanting a small amount of metal ions without changing the original properties of the material substrate [15–17]. In recent years, ion implantation of lanthanum and yttrium has been mainly focused on enhancing the corrosion resistance, high-temperature oxidation resistance, and biocompatibility of pure metals or alloys [18–20]. Xu [21] discovered a significant improvement in the aqueous corrosion resistance of zircaloy-4 by yttrium ion implantation, compared to that of the as-received zircaloy-4. Wang [22] obtained the same conclusion upon exploring the lanthanum ion implantation into a zirconium alloy. Darwin et al. [23] found that ion implantation of lanthanum significantly improved the high-temperature oxidation resistance of an Fe80Cr20 alloy and developed a promising interconnection metal material for planar-type solid oxide fuel cells, namely, Fe80Cr20 60h-La. The co-implantation of yttrium and carbon ions enhanced the wear resistance of H13 steel, which was shown by an increase in the microhardness of 60%–170% and a wear resistance of 2–3 times [24].

However, there have been only a few studies on the application of REs for vacuum carburizing of alloy steel, and the detailed structure and surface properties of the carburizing layer have seldom been analyzed. Thus, to meet the surface property requirements of 20Cr2Ni4A alloy carburized steel, ion-implanted lanthanum and yttrium were used to improve the hardness and surface properties. Due to the large radius of REs, substantial lattice damage and high-density structural defects are produced in the alloy steel matrix during the process of ion implantation. After carburizing, the surface analysis was studied in detail by X-ray photoelectron spectroscopy (XPS), transmission electron microscopy (TEM), X-ray diffraction (XRD), scanning electron microscopy (SEM), and optical microscopy (OM). In addition, the Rockwell and Vickers hardness tests were used to evaluate the hardness of the carburized layer on three different treated samples. Lastly, the catalysis mechanism and the strengthening mechanism of REs during carburization are discussed.

2. Materials and Methods

The parent material of the 20Cr2Ni4A alloy steel (0.17 wt% C, 0.25 wt% Si, 0.45 wt% Mn, 1.45 wt% Cr, 3.48 wt% Ni, and 0.01 wt% Mo) was cut into samples with dimensions of 15 mm × 15 mm × 15 mm. Before ion implantation, these samples were polished with a series of silicon carbide (SiC) abrasive papers with grits of 600# to 2000#, polished with a W3.5 diamond paste and finally ultrasonically cleaned with anhydrous alcohol, for 10 min. Lanthanum and yttrium with 99.99% purity were implanted with a metal vapor vacuum arc (MEVVA) ion source (Shanghai Kingstone Semiconductor Joint Stock Limited Company, Shanghai, China) at the Institute of Low Energy Nuclear Physics in Beijing Normal University (China). The parameters of the ion implantation process are shown in Table 1. Vacuum carburization was carried out after ion implantation of the REs. Vacuum carburization was carried out for 5 hours in a vacuum environment (1×10^{-5} Pa) with a carbon potential of 1.2% and a temperature of 920 °C. To promote the uniform diffusion of carbon atoms, C_2H_2 and N_2 were injected alternately for 25 pulse cycles. Five hours after carburizing at 920 °C, high temperature tempering at

650 °C for 3 hours promoted the precipitation of carbides and the decomposition of retained austenite; then oil quenching at 820 °C was done to improve the hardness and toughness of the material; finally, low temperature tempering was carried out at 180 °C for 3 hours, to eliminate the internal stress of the sample. The whole heat treatment process was completed on the vacuum carburization automatic production line (ECM Technologies, Grenoble, France).

Table 1. Parameters of ion implantation before carburization.

Sample	Implanted with La	Implanted with Y
Vacuum level (Pa)	3.5×10^{-3}	3.5×10^{-3}
Ion energy (Kev)	100	105
Implantation dose	2×10^{17} ions/cm^2	2×10^{17} ions/cm^2
Temperature (°C)	Room temperature	Room temperature
Element	La	Y

The chemical characterization of the implanted surface and the depth profile of the element content distribution of the samples, with depth, were performed by XPS (Thermo Scientific Escalab EI+) (Thermo Fisher Scientific, Waltham, MA, America) using a monochromatic Al radiation source (energy = 55 eV). The acceleration voltage of argon ion etching was 4 kV, and the rate was 5 nm/min. The binding energy of elements was normalized to 284.8 eV of carbon. To illustrate the crystal structure defects of the implantation area after ion implantation, the surfaces of the implanted samples were observed using TEM (JEM-2010) on an instrument equipped with energy dispersive spectroscopy (EDS) (JEOL, Tokyo, Japan). The TEM samples for surface observation by ion implantation were prepared according to the following procedure. Thinning samples were bonded to silicon wafers using epoxy resin glue. Then, the sample were sliced vertically along the plane of the implanted surface and machined to a thickness of about 10 μm. Finally, the samples were ground by Gatan 691 ion thinner (Gatan, Pleasanton, CA, USA) at a small incident angle between 8° and 3°.

The phase composition was studied by XRD (D8 Advance, Bruker AXS, Karlsruhe, Germany, Cu Kα1 = 1.5406 Å) at 40 kV and 40 mA, with an incident angle of 0.5°, after ion implantation and vacuum carburizing. The scanning speed and angle range were 1°/min and 10°–90°, respectively. The penetration depths of X-ray was estimated to be 26 nm at incident angles of 0.5° [25]. Considering the maximum concentration distribution depth of the two implanted ions at 20–25 nm, 0.5° was chosen as the incident angle of the XRD test after carburization. The content of the retained austenite was determined with an X-350A X-ray stress analyzer (ST Stress Technology Co., Ltd., Handan, China). The surface morphology and phase structure of the samples were observed by Optical Microscope (OM) (Olympus, Tokyo, Japan) and a Nova NanoSEM50 environmental SEM (FEI, Hillsboro, OR, USA) equipped with EDS (OXFORD, Oxford, England) was also used. The corrosive fluid for OM comprised an alcohol solution containing 4% nitric acid (volume fraction). To obtain the effective case depth of the carburized layer, according to the GB/T 9450-2005 standard, Vickers hardness testing was used to determine the longitudinal section hardness of the samples under a 1 kgf (HV_1) load, and each hardness value was the average of three measurements. In the Vickers hardness testing, the distance (S) between two adjacent indentation centers should not be less than 2.5 times of the indentation diagonal and the distance difference between successive adjacent indentation centers and part surfaces (a_2–a_1) should not exceed 0.1 mm, as shown in Figure 1. The load time of the apparatus was 15 seconds. The Rockwell hardness of the surface and the core of the samples was measured by a Rockwell hardness tester (Shjingmi, Shanghai, China), and each hardness value was the average of five measurements. For the carburized surface, the depth of Rockwell indenter was about 0.08 mm; for the center, the depth of the Rockwell indenter was about 0.14 mm. The carbon element distribution on the surface of the carburized layer was determined by an electron probe microanalyzer (EPMA-1720, Shimadzu, kyoto, Japan). The test parameters were as follows—acceleration voltage, 10 kV; electron beam current, 200 nA; beam spot, 20 um; and test time, 20s.

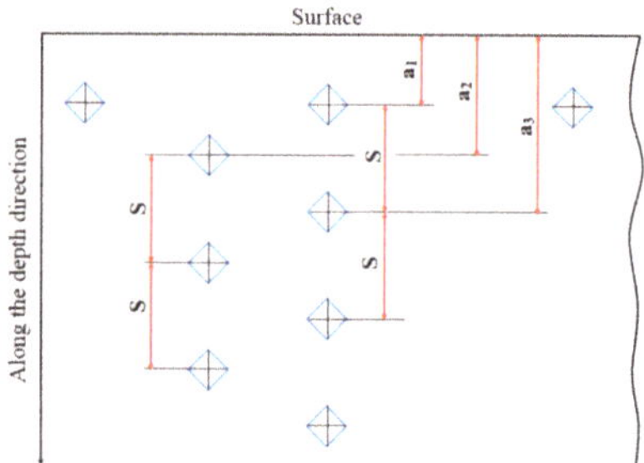

Figure 1. Locations of the hardness indentations.

3. Results

3.1. Calculation of the Implanted Ion Range

The SRIM is commonly used to simulate the ion sputtering process [26]. In this case, the ion range of all implanted ions in the simulation was calculated using dynamic simulation (TRIM) [27]. The impurity distributions of lanthanum and yttrium ions at ionic energies of 100 Kev and 105 Kev were simulated by TRIM. The results are shown in Figures 2 and 3, respectively. As shown in Figures 2a and 3a, each time an implanted ion collided with a target atom, a vacancy (lines of red dots) would be created, which would cause cascade damage (clusters of green dots) of the target atoms in steel. As shown in Figures 2b and 3b, the maximum ionic ranges of lanthanum and yttrium were 50 and 60 nm, respectively, and both had a normal distribution, which was consistent with the XPS results. Figures 2c and 3c show the ionization distribution in which lanthanum and yttrium ions lose their energy in steel samples, and it was observed that the host lattice arrangement of steel was damaged during the implantation of lanthanum and yttrium ions (Figures 2d and 3d).

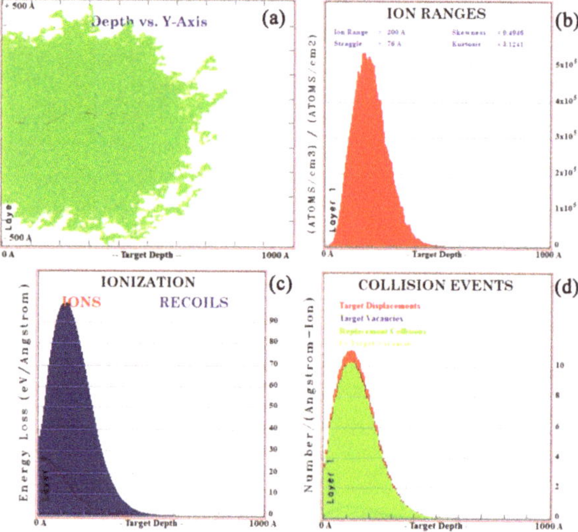

Figure 2. TRIM simulation of lanthanum ion implantation for the (**a**) ion beam pattern, (**b**) ion ranges, (**c**) ionization distribution of ions, and (**d**) collision events.

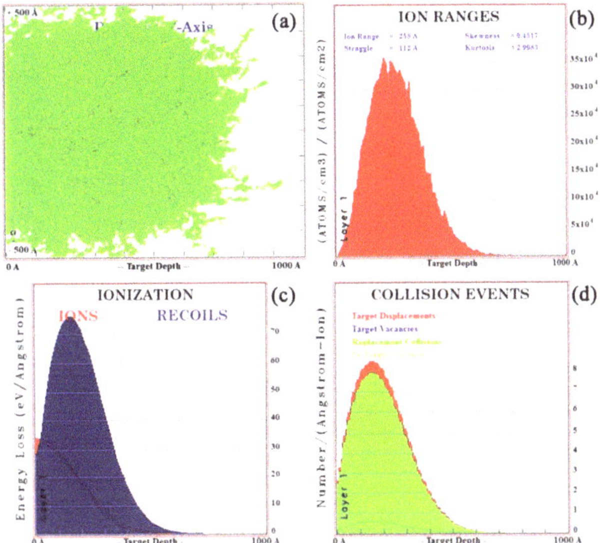

Figure 3. TRIM simulation of yttrium ion implantation for the (**a**) ion beam pattern, (**b**) ion ranges, (**c**) ionization distribution of ions, and (**d**) collision events.

3.2. Chemical Composition and Structure of the Implanted Surface Layer

Figure 4 shows the XPS spectra of the samples after La and Y ion implantation, where contaminants on the surface were removed by Ar^+ etching for 1 minute. Figure 4a,c show the full XPS spectra, which indicate that there are iron, oxygen, carbon, and silicon on the implanted lanthanum and yttrium surfaces, respectively. Compared with the binding energy of the standard absorbed carbon of 284.8 eV, the surface energy of the absorbed carbon in this study was 285.2 eV, which was 0.4 eV higher than that of 284.8 eV. The adjusted XPS data of La and Y are separately illustrated by Figure 4b,d respectively. Figure 4b shows the La 3d surface XPS spectrum. The three peaks correspond to La $3d_{3/2}$ (851.5 eV) and La $3d_{5/2}$ (835.1 eV). The spin-orbit splitting value was 16.4 eV, which clearly suggest a typical La_2O_3 pattern [28]. The phenomenon was also observed by Jin et al. [29]. Figure 4d shows the Y 3d XPS spectrum, and the peaks at 158.0 eV and 155.9 eV correspond to the Y–Y bond, and the peak at 157.4 eV corresponds to Y_2O_3. Consequently, this result illustrate that lanthanum mainly exists as oxides, while yttrium exists as oxides and metallic Y in the RE implanted layer.

Figure 5 shows the depth profiles of the element content distribution of lanthanum and yttrium ion implantation samples from the XPS tests. Figure 5a,b show that the concentration of lanthanum and yttrium in the ion implantation layer present a normal distribution trend, and the peak concentration distributions are at depths of approximately 15 and 25 nm, with concentrations of 22.9 wt% and 21.2 wt%, respectively.

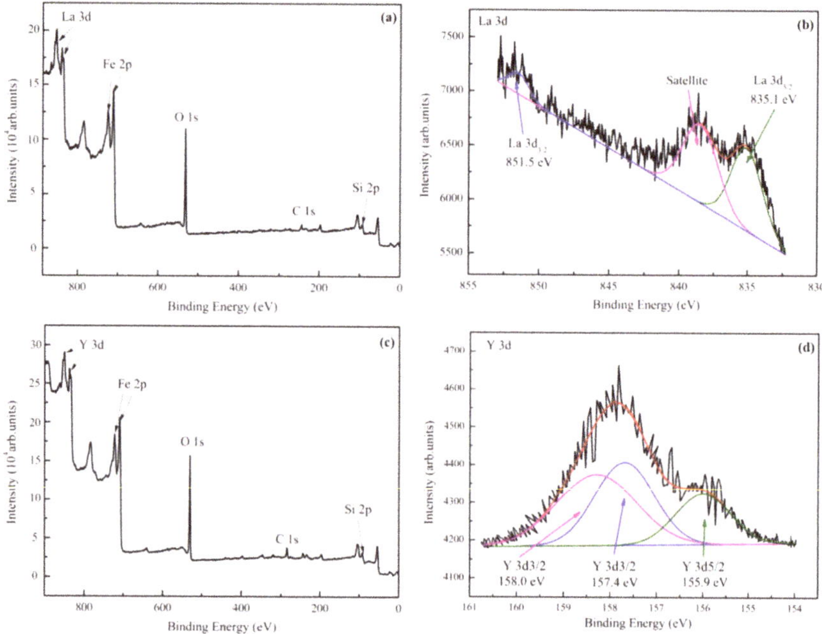

Figure 4. XPS spectra of sample surfaces—(**a**) survey spectrum and (**b**) La 3d spectrum of implanted La; (**c**) survey spectrum and (**d**) Y 3d spectrum of implanted Y.

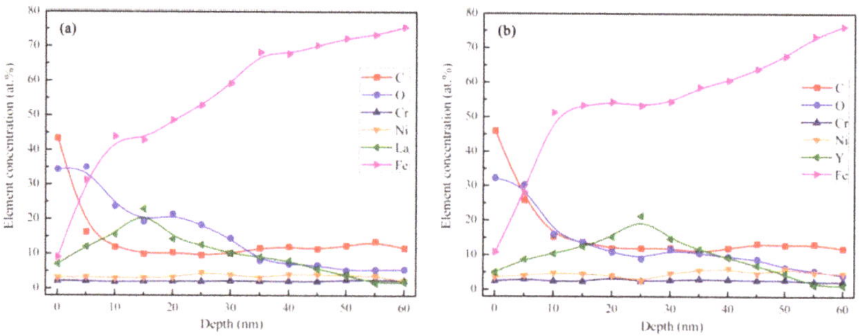

Figure 5. Depth profiles of element content distribution obtained by XPS—(**a**) lanthanum implantation result and (**b**) yttrium implantation result.

As shown in Figure 6, due to the larger radius of rare earth and the internal stress introduced by ion implantation [30], all three diffraction peaks of α-Fe shift to the left and the peak intensity increased, which meant that the lattice distortion occurred on the surface of the implanted matrix, according to the Bragg equation [31].

Figure 7 shows the TEM image of the surface layer of the samples, with and without ion implantation. It can be observed from Figure 7a,b that that there are some crystal defects in the 20Cr2Ni4A matrix, such as dislocation entanglement and dislocation grid, which form a stable defect network. The RE ion implantation caused lattice damage to the body-centered cubic (BCC) Fe matrix. From the observation of Figure 7c–f, the matrix implanted with rare earth ions forms high density dislocation entanglement compared to that of the non-implanted sample. Especially after yttrium ion implantation, the matrix was almost full of high-density dislocations, which accumulated and tangled [30].

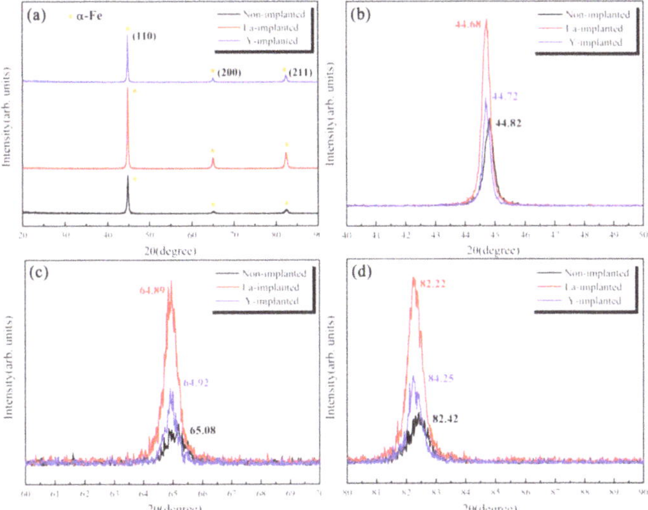

Figure 6. XRD patterns of the implanted surface with and without rare earth (RE) implantation—(**a**) XRD spectra; (**b**) diffraction angle of the (110) peak; (**c**) diffraction angle of the (200) peak; and (**d**) diffraction angle of the (211) peak.

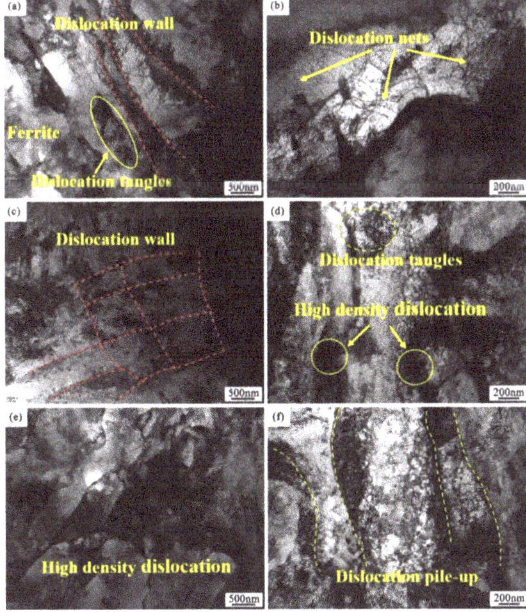

Figure 7. TEM observation of non-carburized samples after ion implantation—(**a**,**b**) non-implanted sample; (**c**,**d**) lanthanum-implanted sample; and (**e**,**f**) yttrium-implanted sample.

3.3. Phase and Microstructure of the Carburized Layer After Ion Implantation

The phase on the surface of the carburized samples was measured by XRD, as shown in Figure 8. The carburized layers of the three samples are composed of a martensite phase with a body-centered cubic (BCC) structure and retained austenite with a face-centered cubic (FCC) structure. Due to the limitation of the phase detection spatial resolution with XRD, there are no diffraction peaks from lanthanum and yttrium. However, it was found that the strongest diffraction peaks from the lanthanum and yttrium-implanted samples shift to the left by 0.26 and 0.20 degrees, respectively, compared with that of the non-implanted sample. The other two peaks of martensite migrated to the left at the same time. In addition, the intensity of all crystal planes after lanthanum and yttrium implantation were stronger than those from the non-implanted sample. X-ray stress analyses also revealed that the content of the retained austenite in the carburized layer decreased slightly with the implantation of RE ions, as shown in Table 2.

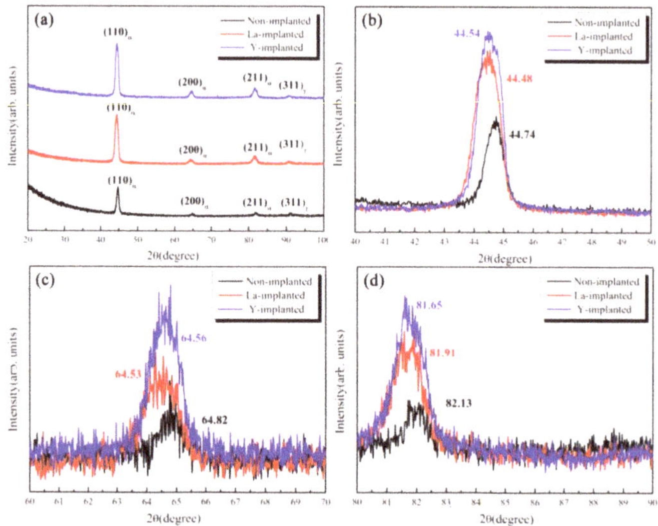

Figure 8. XRD spectra of the carburized layers of the three samples—(**a**) XRD spectra; (**b**) diffraction angle of the (110) peak; (**c**) diffraction angle of the (200) peak; and (**d**) diffraction angle of the (211) peak.

Table 2. Retained austenite content in the carburized surface layer of the three samples by XRD.

Sample	Non-Implanted	Lanthanum-Implanted	Yttrium-Implanted
Retained austenite	16.7%	15.6%	14.2%

In this section, the microstructures of the carburized layers of different samples carburized in vacuum for 5 hours at 920 °C are discussed. Figure 9a,c,e show the layers from conventional carburizing and carburizing with REs lanthanum and yttrium, respectively. Figure 9a shows that the microstructure of the carburized layer was composed of a martensite matrix, and granular carbides were dispersed in it and retained austenite, which were not transformed into martensite. However, ultrafine acicular martensite and fine dispersed grain carbides were obtained in the carburized layer after the ion implantation of lanthanum and yttrium, as shown in Figure 9c,e, respectively. Figure 9b,d,f show that the core microstructures of the three samples were tempered martensite and free ferrite. According to the GB/T 25744-2010 metallographic standard, the core microstructures of the non-implanted sample were equivalent to grade 4, while those of the samples with RE ion implantation were equivalent to grade 3.

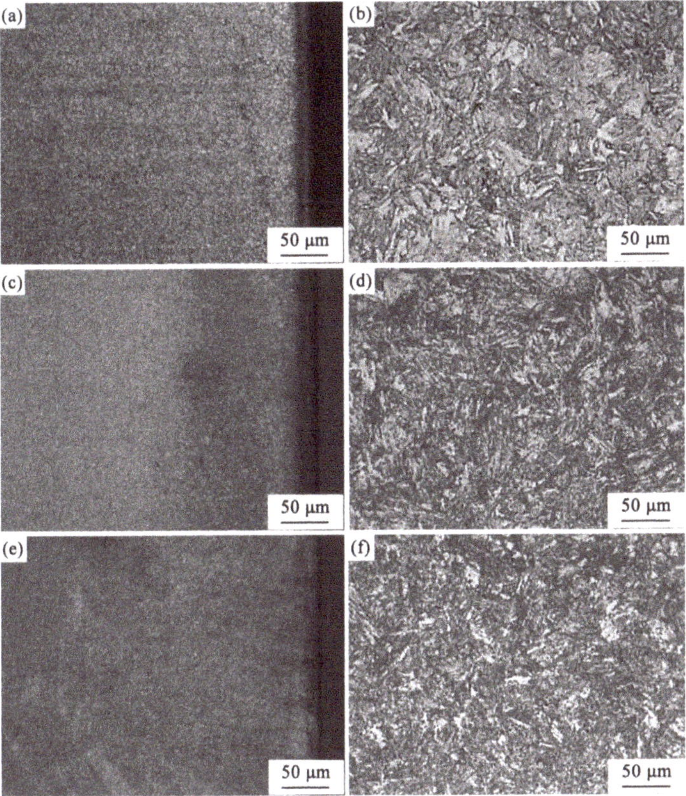

Figure 9. Microstructure of the carburized layer on the three samples—(**a**) carburized surface layer without RE; (**b**) core after carburizing without RE; (**c**) carburized surface layer after La implantation; (**d**) core after La implantation; (**e**) carburized surface layer after Y implantation; and (**f**) core after Y implantation.

In addition, the SEM images of the carbides in the carburized layer and their corresponding size and distribution histograms are shown in Figure 10. Figure 10a shows that the majority of the carbides are granular and unevenly distributed in the interstices between the martensites and are accompanied by large diameter bar carbides. However, compared to that in the non-implanted layer, the carbide distribution in the carburized layer of the sample after ion implantation was finer and more uniform. Among the non-implanted and implanted layers, the carbides implanted with yttrium were the finest. The EDS measurements of the carbide particles in the three samples showed that the carbides were mainly composed of carbon, iron, chromium, nickel, and manganese, while lanthanum and yttrium were found in the carbides in the ion-implanted samples. The size distribution of the carbide particles was estimated by using Nano Measurer 1.2 software and Gaussian fitting (v1.2, Wan An Intelligent Technology, Wuhan, China). At least 150 particles were counted from each SEM image. The carbide particles in SEM photos were labeled and statistical reports were derived [32]. The average diameter of carbides on the surface of the non-implanted samples was 0.35 μm. Among them, 81.2% of the carbides were within 0.6 μm in diameter, and the length of a portion of the large strip carbides was more than 5 μm. However, the average diameters of the carbides on the surface of the carburized layer after ion implantation of lanthanum and yttrium were 0.25 and 0.17 μm, respectively. The diameters of most carbides were from 0.1 to 0.3 μm. The results showed that ion implantation of the REs reduced the particle size of carbides on the carburized layer surface. The effect of yttrium ion implantation was better than that of the lanthanum ion implantation because yttrium implantation resulted in a minimum particle size of 0.06 µm and a maximum particle size of 0.36 µm. In other words, RE ion implantation pretreatment played an important role in refining the structure of the carburized layer and promoting the dispersion of the fine carbide precipitates.

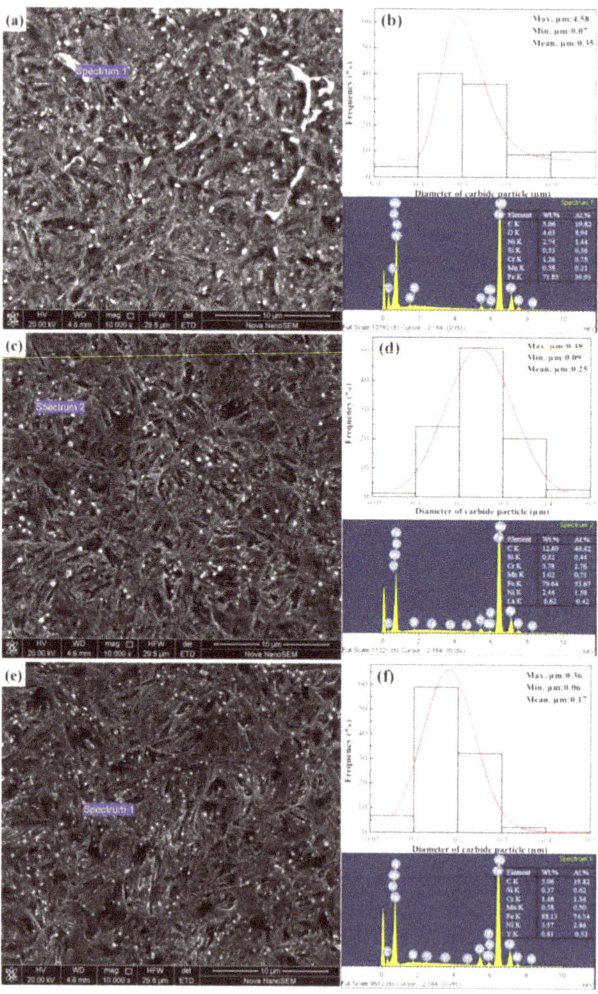

Figure 10. Diameter of the granular carbides in the carburized surface layer—(**a**) carbide morphology and energy dispersive spectroscopy (EDS) results without RE and (**b**) its carbide diameter distribution; (**c**) carbide morphology and EDS results with La implantation and (**d**) its carbide diameter distribution; (**e**) carbide morphology and EDS results with Y implantation, and (**f**) its carbide diameter distribution.

3.4. Hardness Distribution and Effective Hardening Depth of the Carburized Layer

Figure 11 shows the results of the Rockwell hardness measurements. The surface Rockwell hardness of the non-implanted sample was 58.3 HRC, and the hardness of RE-implanted samples was higher than the previous value, especially samples with the yttrium addition. After vacuum carburization, the Rockwell hardness of the yttrium-implanted sample increased to 61.8 HRC at the subsurface layer; this value was higher than that of the lanthanum-implanted sample (60.5 HRC). In contrast, the core hardness value did not change significantly. The smaller error bar value indicated that there was no significant difference in the range of surface hardness.

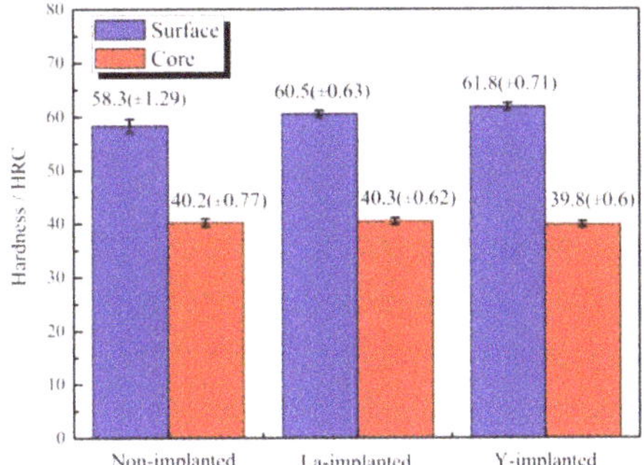

Figure 11. Rockwell hardness of the carburized layer and the core of the three samples.

Figure 12 shows the microhardness results of the samples after carburization, with and without the RE ions. The fluctuation of the error bars were not more than 25 HV_1. It could be seen that the microhardness of the carburized layer after RE ion implantation was higher than that treated by conventional vacuum carburization at 920 °C, and the change in the hardness gradient was minor. The maximum hardness of the carburized layer after ion implantation of lanthanum and yttrium occurred at 0.1 mm and 0.2 mm from the surface and was 805 HV_1 and 822 HV_1, respectively, which was higher than the value of 778 HV_1 at 0.2 mm from the surface, after conventional carburizing. Osman Asi suggested that the enhancement in hardness was mainly caused by the diffusion of carbon atoms [33]. According to the GBT 9450-2005 standard, the effective hardening depths of the non-implanted, La-implanted, and Y-implanted samples were 1.36 mm, 1.44 mm, and 1.47 mm, respectively, as shown by the arrows in Figure 12. RE ion implantation increased the hardness of the carburized layer of the 20Cr2Ni4A steel but did not cause a significant increase in the depth of the effective hardened layer. Hence, the results showed that the microhardness of the carburized layer obtained by vacuum carburizing with yttrium was the highest, and the change was the most uniform among the samples considered herein.

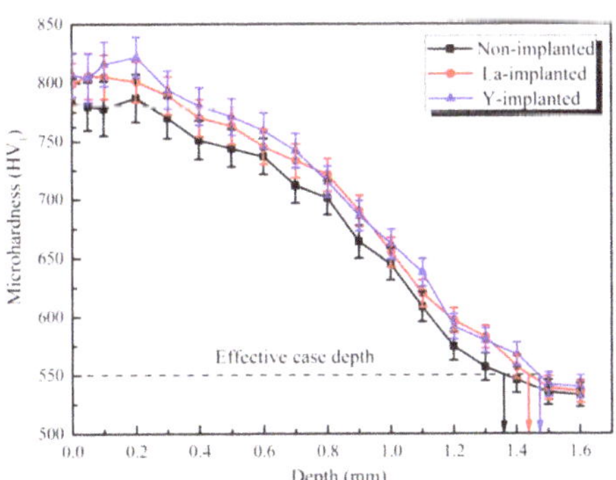

Figure 12. The microhardness distribution of the carburized layer along the depth direction.

3.5. Calculation of the Carbon Diffusion Coefficient

As can be seen from Figure 13, the carbon concentration on the carburized surface without ion implantation was about 0.84%, and the carbon concentration varied unevenly. After ion implantation, the carbon content on the carburized surface reached more than 0.9%, and the carbon concentration gradient changed more gently. The hardness gradient distribution of the three carburized specimens was approximately the same as that of the carbon element distribution measured by the EPMA.

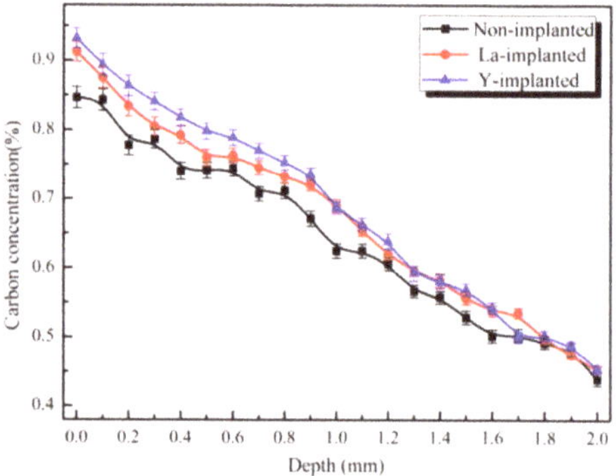

Figure 13. Carbon concentration along the depth of the carburized layers.

During the vacuum carburization process, acetylene gas was the carbon source. The diffusion of carbon atoms followed Fick's second law and was considered a one-dimensional plane diffusion phenomenon. If the diffusion coefficient did not change with the carbon concentration (or average diffusion coefficient), the carbon concentration distribution in the infiltration process satisfied Equation (1) [12]:

$$\frac{\partial c(x,t)}{\partial_t} = D\frac{\partial^2 c(x,t)}{\partial x^2} \tag{1}$$

where $c(x, t)$ is the volume concentration of carbon, x is the distance from any point in the sample to the surface of the sample, t is the diffusion time, and D is the carbon diffusion coefficient.

With the extension of the carburizing time, the carbon concentration c_s on the steel surface gradually increased from the original carbon content c_0 to a constant level and was in equilibrium with the carbon potential c_p of the furnace gas, which indicated that it had entered the stage of carbon diffusion. At the beginning of the carburization process, the initial and boundary conditions of Equation (1) were marked as $c_0 = (x, 0)$ and $c_s = (0, t)$. The solution of Fick's second law, which provided the curve of the carbon diffusion concentration, was as Equation (2):

$$\frac{c_x - c_0}{c_s - c_0} = 1 - erf\left(x/2\sqrt{Dt}\right) \tag{2}$$

where $erf\,(x/2\sqrt{Dt})$ is the error function. When the specified value of c_x was higher than $c_{0.35\%}$, the obtained depth of the hardened layer could be described as Equation (3):

$$x = 2K\sqrt{Dt} \tag{3}$$

where K is a constant (constant carbon potential).

Therefore, based on the depth of the carburized layer shown in Figure 12, when the carburizing temperature and time were the same, the diffusion coefficient relation of the carburized layer with or without RE ions could be obtained by Equation (3), as shown in Equations (4) and (5):

$$1.44/1.36 = \sqrt{D^{La}_{920°C}/D_{920°C}} \tag{4}$$

$$1.47/1.36 = \sqrt{D^{Y}_{920°C}/D_{920°C}} \tag{5}$$

where $D_{920°}$, $D_{920°}^{La}$, and $D_{920°}^{Y}$ are the carbon diffusion coefficients of 20Cr2Ni4A steel without ion implantation and with lanthanum or yttrium implantation before vacuum carburizing, respectively. Finally, the relations were shown in Equations (6) and (7):

$$D_{920°C}^{La} / 1.12 D_{920°C} \tag{6}$$

$$D_{920°C}^{Y} / 1.17 D_{920°C} \tag{7}$$

In short, the carbon diffusion coefficients after lanthanum and yttrium implantation were 1.12 and 1.17 times higher, respectively, than those of the non-implanted samples when the vacuum carburizing temperature was 920 °C and the carbon potential was 1.2%. This was consistent with the results of the hardness gradient distribution and the carbon concentration distribution of the carburized layer.

4. Discussion

4.1. Effect of RE on Carbon Diffusion

As shown in Figures 4 and 5, after ion implantation, the 20Cr2Ni4A matrix formed an injection layer containing the RE, in which the RE atoms existed in the form of oxides and solid solution atoms. The results of the TEM analysis in Figure 7 showed that RE ion implantation caused substantial damage to the matrix lattice.

Due to the characteristics of the electron cloud of RE element shells and its atomic size effect, a series of cascade collisions occurred between ion-implanted RE atoms and substrate atoms, resulting in a large number of dislocations and damage in the matrix lattice [34,35]. The estimation of mean ion energy transferred to the target were reasonably evaluated by SRIM. The average primary knock-on atom energy of the La and Y ions were 94.6 and 96.8 Kev/ion during the ion implantation, respectively. As shown in Figure 14, the maximum electronic stopping powers of La and Y ions along the depth direction were 537.8 and 370.3 Kev/Å, respectively (electron stopping power is the blocking effect of nucleus and electrons in the target matrix on the implanted ions during ion implantation [34]). A schematic diagram is shown in Figure 15. Subsequently, the disorder degree of the matrix atoms, the crystal interface area and the crystal structure defects increased, which became the channels for carbon diffusion during carburizing and played an extremely important role in improving the carbon diffusion coefficient [11].

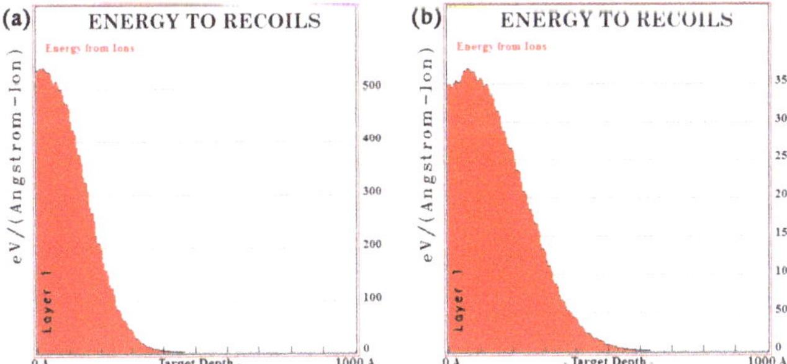

Figure 14. SRIM simulation of mean ion energy during ion implantation of (**a**) La and (**b**) Y into steel.

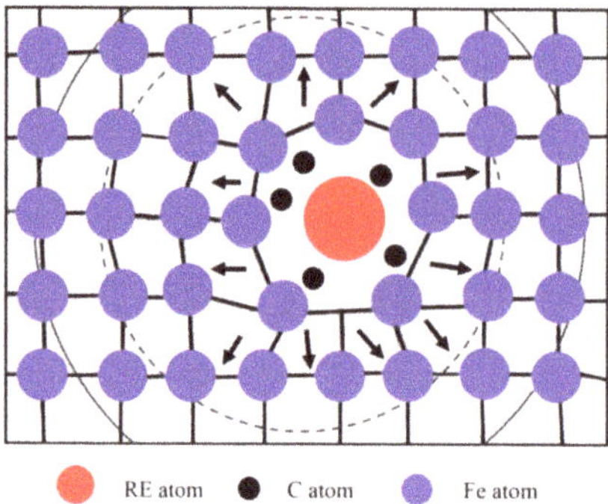

Figure 15. Physical model of an RE solid solution atom in a crystal face in an Fe lattice.

When solid solution RE atoms expand the lattice defects in the matrix, the distorted lattice region of the surrounding iron atoms becomes a trap for carbon atoms during carburization. Carbon atoms segregate into the voids of the distortion region [36], resulting in the formation of a nanoscale Cottrell atmosphere with the RE as the core carbon atom [2], as shown in Figure 16. Moreover, as shown in Figure 16a, carbon uniformly diffuses in the volume from the outside to the inside, in accordance with the concentration gradient in the austenite during conventional carburizing. The grain boundary diffusion was faster than that in the grains, and the finer the grain was, the faster was the diffusion rate. However, once the Cottrell atmosphere was formed during carburizing with RE ions, the diffusion mode changed from uniform diffusion to non-uniform diffusion [37]. As shown in Figure 16b, the Cottrell atmosphere also acted as an accelerator for diffusion [38]. First, the concentration of carbon atoms in the air mass was much higher than that in the matrix, resulting in a high concentration difference. Then, the larger the density of the air mass and the higher the average carbon concentration, the larger was the diffusion flux, such that the diffusion velocity and diffusion flux were significantly increased. Finally, the mechanism of carbon diffusion during RE carburization could be considered to be a result of jumping the short-circuit diffusion when the carbon atoms at the top of the last gas mass jumped to the next one. It could be concluded that the lattice damage and associated Cottrell atmosphere produced by the RE ion implantation changed the diffusion mode of the carbon atoms during the conventional carburizing processes and increased the diffusion coefficient of the carbon atoms.

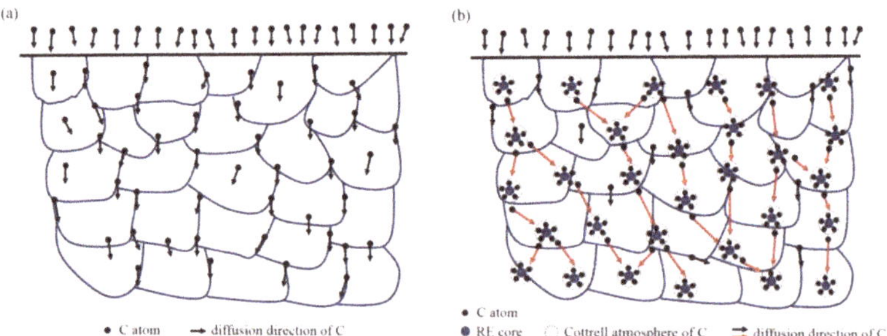

Figure 16. Diffusion models of (**a**) conventional carburizing and the (**b**) RE carburizing processes.

4.2. Effect of REs on the Microstructure and Hardness of the Carburized Layer

According to the differences in the microstructure, phase, and hardness of the carburized layers between the non-implanted and implanted samples, as determined by the OM, SEM, and the XRD, it could be concluded that RE ion implantation can improve the carburized layer [10]. From Figure 8, the diffraction peaks of the ion-implanted lanthanum and yttrium specimens shift to the left, and the intensity of the peaks increased after vacuum carburizing. From the Bragg equation, the smaller the diffraction angle, the larger the lattice plane distance. Due to the large radius of RE atoms, the iron lattice in steel is distorted [39]. Moreover, the (110) α diffraction peak showed and obvious increase, which indicated that the preferred orientation of cryptocrystalline martensite was (110) α after RE ion implantation, and its content also increased. From the results in Figures 9 and 10, it could be concluded that the structure of the carburized layer after ion implantation was obviously refined, the content of the cryptocrystalline martensite was increased, and the minimum size of the surface carbide was 0.17 μm. Figure 11 shows that the surface hardness increased after RE ion implantation.

These results could be explained according to the following two aspects. First, the RE ion implantation pretreatment could change the martensite transformation mode. Since martensite has an explosive shear growth, the first martensite could not pass through the austenite large-angle grain boundary, and it was impossible to cut through the carbides. During the conventional carburization process, substantial amounts of carbides do not exist in the austenite crystal, so the martensite tends to be coarse. Carbides precipitated from austenite crystals are often required during RE carburizing. In the presence of these carbides, the martensite shear is blocked and the martensite is forced to become superfine [10]. Moreover, the quenching performance was also improved. Figure 17 shows the martensite transformation patterns of the two carburizing methods. Figure 17a shows the transformation mode of austenite to martensite during conventional carburizing and quenching, and it can be seen that plate martensite needles often appeared. Figure 17b shows that during the carburizing and quenching processes with REs, the Cottrell atmosphere with RE ions as the core becomes the nucleation core for a carbide, which results in the precipitation of fine dispersed carbides on the surface, during carburization and makes the martensite superfine, as shown in Figure 10c,e. The fine, dispersed, spherical carbides were embedded in the martensite matrix, which inhibited the growth of the austenite grains and improved the hardness and wear resistance of the steel.

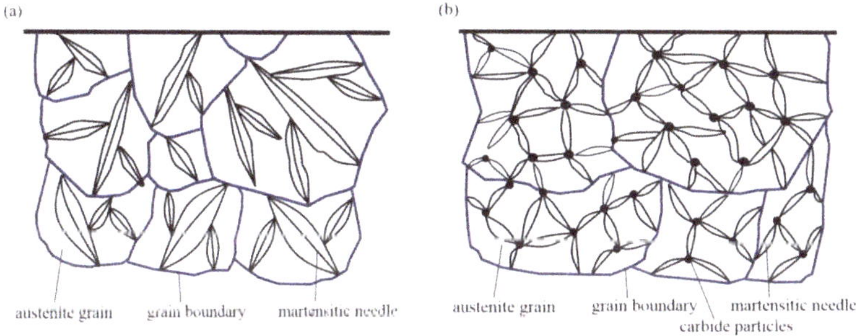

Figure 17. Transformation models of martensite in the grain interior for (**a**) conventional carburization and the (**b**) RE carburization processes.

In addition, RE ion implantation pretreatment could dramatically refine the microstructure of the carburized layer. The undercooled austenite before quenching was uniform and stable during conventional carburization. RE carburization results in a typical non-uniform structure, as its stability is very poor due to the carbide nucleus [10]. Without undercooling, the carbon concentration around the carbide increased to the highest level and decreased far away from the carbide. The carbon concentration periodically changed according to the distance between the carbides. With an increase in the undercooling before quenching, the carbon saturation in austenite around the carbide increased sharply, causing the carbon atoms in the austenite to remain around the carbide. With the uphill diffusion process from the austenite to the carbide, that is, from a low concentration to a high concentration of carbon, the carbon in the austenite became depleted. After this transformation, the superfine martensite was formed. The resistance of the superfine martensite to fatigue crack initiation and propagation was greatly improved, so the microstructure contributed to increasing the service life of the different parts.

5. Conclusions

1. La and Y ion implantation can result in the appearance of La_2O_3, Y_2O_3 and Y (0). In the implanted layer, the maximum concentration of La and Y occurred at 15 and 20 nm, respectively, and the concentration of both ions showed a normal distribution.

2. During the carburization process, the high density of dislocation defects and the Cottrell atmosphere that formed on the surface of the steel by the RE ion implantation became the diffusion channels for the carbon atoms. During the carburization process at 920 °C and with a 1.2% carbon potential, the La ion implantation increased and the carbon diffusion coefficient was 1.12 times higher than that of the non-implanted steel, which was beneficial. Y ion implantation increased the carbon diffusion coefficient, which was 1.17 times higher than that of the non-implanted steel, which was also beneficial. The effective hardening depth of the carburized layer increased by 0.08 mm and 0.11 mm after the La and Y ion implantation, respectively.

3. RE ion implantation pretreatment could change the transformation mode of the martensite and could significantly refine the microstructure of the carburized layer. Most of the fine dispersed carbides have diameters ranging from 0.1 mm to 0.2 mm and the maximum diameters were less than 0.32 mm. The maximum diameter and minimum diameter of the carbides on the surface of the carburized layer after Y ion implantation were 0.36 mm and 0.06 mm, respectively. The surface hardness of the carburized layer after ion implantation of lanthanum and yttrium was 60.5 HRC and 61.8 HRC, respectively. The yttrium-implanted carburized layer had a higher surface hardness than the non-implanted carburized layer, showing an increase of 3.5 HRC. Therefore, the results showed that compared to lanthanum, yttrium has a better strengthening effect.

Author Contributions: Conceptualization, C.L. and Z.X.; methodology, C.L.; software, G.L. and W.G.; validation, C.L. and Z.X.; formal analysis, G.L.; investigation, C.L.; resources, H.L.; data curation, C.L.; writing—original draft preparation, C.L.; writing—review and editing, C.L.; Y.H.; visualization, H.W.; supervision, G.L.; project administration, Z.X.; funding acquisition, H.W.

Funding: This work was financially supported by the National Natural Science Foundation of China (51535011, 51775554) and the 973 Project (61328304).

Acknowledgments: The author is very grateful to Shulan Zhang and Ying Chen from Central Iron & Steel Research Institute for their technical assistance.

Conflicts of Interest: The authors declare no conflict of interest.

References

1. Wang, F.F.; Zhou, C.G.; Zheng, L.J.; Zhang, H. Improvement of the corrosion and tribological properties of CSS-42L aerospace bearing steel using carbon ion implantation. *Appl. Surf. Sci.* **2017**, *392*, 305–311. [CrossRef]

2. Dong, M.L.; Cui, X.F.; Jin, G.; Wang, H.D.; Cai, Z.B.; Song, S.Q. Improved microstructure and properties of 12Cr2Ni4A alloy steel by vacuum carburization and Ti + N co-implantation. *Appl. Surf. Sci.* **2018**, *440*, 660–668. [CrossRef]

3. Paulson, N.R.; Golmohammadi, Z.; Walvekar, A.A.; Sadeghi, F.; Mistry, K. Rolling contact fatigue in refurbished case carburized bearings. *Tribol. Int.* **2017**, *115*, 348–364. [CrossRef]

4. Walvekar, A.A.; Sadeghi, F. Rolling contact fatigue of case carburized steels. *Int. J. Fatigue* **2017**, *95*, 264–281. [CrossRef]

5. Guo, Y.B.; Zhang, Z.; Zhang, S.W. Advances in the application of biomimetic surface engineering in the oil and gas industry. *Friction* **2019**, *7*, 289–306. [CrossRef]

6. Zajusz, M.; Tkacz-Śmiech, K.; Danielewski, M. Modeling of vacuum pulse carburizing of steel. *Surf. Coat. Technol.* **2014**, *258*, 646–651. [CrossRef]

7. Dychtoń, K.; Rokicki, P.; Nowotnik, A.; Drajewicz, M.; Sieniawski, J. Process Temperature Effect on Surface Layer of Vacuum Carburized Low-Alloy Steel Gears. *Solid State Phenom.* **2015**, *227*, 425–428. [CrossRef]

8. Wei, S.; Wang, G.; Zhao, X.; Zhang, X.; Rong, Y. Experimental Study on Vacuum Carburizing Process for Low-Carbon Alloy Steel. *J. Mater. Eng. Perform.* **2014**, *23*, 545–550. [CrossRef]

9. Li, G.; Hua, J. The influence of additive rare earths on ion carburization. *Surf. Coat. Technol.* **1993**, *59*, 117–120. [CrossRef]

10. Yan, M.F.; Liu, Z.R. Study on microstructure and microhardness in surface layer of 20CrMnTi steel carburised at 880 °C with and without RE. *Mater. Chem. Phys.* **2001**, *72*, 97–100. [CrossRef]

11. Yan, M.F.; Pan, W.; Bell, T.; Liu, Z. Effect of rare earth catalyst on carburizing kinetics in a sealed quench furnace with endothermic atmosphere. *Appl. Surf. Sci.* **2001**, *173*, 91–94. [CrossRef]

12. Yan, M.F.; Liu, Z.R.; Bell, T. Effect of Rare Earths on Diffusion Coefficient and Transfer Coefficient of Carbon during Carburizing. *J. Rare Earth* **2001**, *19*, 122–124.

13. Dong, M.L.; Cui, X.F.; Zhang, Y.H.; Jin, G.; Yue, C.W.; Zhao, X.; Cai, Z.B.; Xu, B.S. Vacuum carburization of 12Cr2Ni4A low carbon alloy steel with lanthanum and cerium ion implantation. *J. Rare Earth* **2017**, *35*, 1164–1170. [CrossRef]

14. Chen, X.H.; Soveja, A.; Chaussumier, M.; Zhang, P.Z.; Wei, D.B.; Ding, F. Effect of MEVVA ion implantation on fatigue properties of TC18 titanium alloy. *Surf. Coat. Technol.* **2018**, *344*, 572–578. [CrossRef]

15. Panin, S.V.; Vlasov, I.V.; Sergeev, V.P.; Maruschak, P.O.; Sunder, R.; Ovechkin, B.B. Fatigue life improvement of 12Cr1MoV steel by irradiation with Zr + ion beam. *Int. J. Fatigue* **2015**, *76*, 3–10. [CrossRef]

16. Chen, X.H.; Zhang, P.Z.; Wei, D.B.; Huang, X.; Adriana, S.; Michel, C.; Ding, F.; Li, F.K. Structures and properties of Ti-5Al-5Mo-5V-1Cr-1Fe after Nb implantation. *Surf. Coat. Technol.* **2019**, *358*, 676–687. [CrossRef]

17. Wang, S.X.; Li, C.; Xiong, B.J.; Tian, X.B.; Yang, S.Q. Surface modification of hard alloy by Y ion implantation under different atmosphere. *Appl. Surf. Sci.* **2011**, *257*, 5826–5830. [CrossRef]

18. Zhu, S.F.; Huang, N.; Shu, H.; Wu, Y.P.; Xu, L. Corrosion resistance and blood compatibility of lanthanum ion implanted pure iron by MEVVA. *Appl. Surf. Sci.* **2009**, *256*, 99–104. [CrossRef]

19. Wang, X.M.; Zeng, X.Q.; Wu, G.S.; Yao, S.S. Yttrium ion implantation on the surface properties of magnesium. *Appl. Surf. Sci.* **2006**, *253*, 2437–2442. [CrossRef]

20. Bennett, M.J.; Tuson, A.T. Improved high temperature oxidation behaviour of alloys by ion implantation. *Mat. Sci. Eng. A Struct.* **1989**, *116*, 79–87. [CrossRef]

21. Xu, J.; Bai, X.D.; Jin, A.; Fan, Y.D. Effect of yttrium ion implantation on aqueous corrosion resistance of zircaloy-4. *J. Mater. Sci. Lett.* **2000**, *19*, 1633–1635. [CrossRef]

22. Qian, W.; Bai, X.D.; Liu, X.Y.; Zhao, X. Studies on the corrosion behavior of lanthanum-implanted zircaloy. *J. Mater. Sci.* **2005**, *40*, 475–479. [CrossRef]

23. Sebayang, D.; Khaerudini, D.S.; Saryanto, H.; Hasan, S.; Othman, M.A.; Untoro, P. Oxidation Resistance of Fe80Cr20 Alloys Treated by Rare Earth Element Ion Implantation. In Proceedings of the American Institute of Physics Conference Series; American Institute of Physics: Melville, NY, USA, 2011; Volume 1394, pp. 90–102.

24. Zhang, T.H.; Xie, J.D.; Ji, C.Z.; Chen, J.; Hong, X.; Li, J.; Sun, G.R.; Zhang, H.X. Influence of the structure of implanted steel with Y, Y + C and Y + Cr on the behaviors of wear, oxidation and corrosion resistance. *Surf. Coat. Technol.* **1995**, *72*, 93–98. [CrossRef]

25. Dudognon, J.; Vayer, M.; Pineau, A.; Erre, R. Mo and Ag ion implantation in austenitic, ferritic and duplex stainless steels: A comparative study. *Surf. Coat. Technol.* **2008**, *203*, 180–185. [CrossRef]

26. Shulga, V.I. Note on the artefacts in SRIM simulation of sputtering. *Appl. Surf. Sci.* **2018**, *439*, 456–461. [CrossRef]

27. Chrobak, Ł.; Maliński, M. Properties of silicon implanted with Fe+, Ge+, Mn+ ions investigated using a frequency contactless modulated free-carrier absorption technique. *Opt. Mater.* **2018**, *86*, 484–491. [CrossRef]

28. Peng, D.Q.; Bai, X.D.; Chen, B.S. Surface analysis and corrosion behavior of zirconium samples implanted with yttrium and lanthanum. *Surf. Coat. Technol.* **2005**, *190*, 440–447. [CrossRef]

29. Jin, H.M.; Zhou, X.W.; Zhang, L.N. Effects of lanthanum ion-implantation on microstructure of oxide film formed on Co-Cr alloy. *J. Rare Earth* **2008**, *26*, 406–409. [CrossRef]

30. Sharkeev, Y.P.; Kozlov, E.V. The long-range effect in ion implanted metallic materials: Dislocation structures, properties, stresses, mechanisms. *Surf. Coat. Technol.* **2002**, *158*, 219–224. [CrossRef]

31. Dai, M.Y.; Li, C.Y.; Hu, J. The enhancement effect and kinetics of rare earth assisted salt bath nitriding. *J. Alloys Compd.* **2016**, *688*, 350–356. [CrossRef]

32. Yin, Y.G.; Shen, M.H.; Tan, Z.Q.; Yu, S.J.; Liu, J.F.; Jiang, G.B. Particle coating-dependent interaction of molecular weight fractionated natural organic matter: Impacts on the aggregation of silver nanoparticles. *Environ. Sci. Technol.* **2015**, *49*, 6581–6589. [CrossRef] [PubMed]

33. Asi, O.; Can, A.Ç.; Pineault, J.; Belassel, M. The effect of high temperature gas carburizing on bending fatigue strength of SAE 8620 steel. *Mater. Des.* **2009**, *30*, 1792–1797. [CrossRef]

34. Ziegler, J.F.; Ziegler, M.D.; Biersack, J.P. SRIM—The stopping and range of ions in matter (2010). *Nucl. Instrum. Meth. B* **2010**, *268*, 1818–1823. [CrossRef]

35. Cutroneo, M.; Torrisi, L.; Havranek, V.; Mackova, A.; Malinsky, P.; Torrisi, A.; Stammers, J.; Sofer, Z.; Silipigni, L.; Fazio, B.; et al. Characterization of graphene oxide film by implantation of low energy copper ions. *Nucl. Instrum. Meth. B* **2019**. [CrossRef]

36. Waseda, O.; Veiga, R.G.; Morthomas, J.; Chantrenne, P.; Becquart, C.S.; Ribeiro, F.; Jelea, A.; Goldenstein, H.; Perez, M. Formation of carbon Cottrell atmospheres and their effect on the stress field around an edge dislocation. *Scr. Mater.* **2017**, *129*, 16–19. [CrossRef]

37. Liu, Z.R.; Yan, M.F.; Luo, Q.; Zheng, T.Q.; Chen, Y.J. Diffusion and micro-alloying mechanism of rare earth(RE) and carbon/nitrogen atoms permeated into surface layer of steel during RE-carburizing,-nitriding and-nitrocarburizing. *Mater. Heat Treat.* **2011**, *32*, 121–129.

38. Wang, X.A.; Yan, M.F.; Liu, R.L.; Zhang, Y.X. Effect of rare earth addition on microstructure and corrosion behavior of plasma nitrocarburized M50NiL steel. *J. Rare Earth* **2016**, *34*, 1148–1155. [CrossRef]

39. Liu, D.R.; Qi, Z.; Qin, Z.B.; Qin, L.; Wu, Z.; Liu, L. Tribological performance of surfaces enhanced by texturing and nitrogen implantation. *Appl. Surf. Sci.* **2016**, *363*, 161–167. [CrossRef]

40. Yuan, Z.X.; Yu, Z.S.; Tan, P.; Song, S.H. Effect of rare earths on the carburization of steel. *Mater. Sci. Eng. A Struct.* **1999**, *267*, 162–166. [CrossRef]

 *materials*

Review

Correlation between Microstructural Alteration, Mechanical Properties and Manufacturability after Cryogenic Treatment: A Review

Abbas Razavykia, Cristiana Delprete and Paolo Baldissera *

Department of Mechanical and Aerospace Engineering, Politecnico di Torino, 10138 Torino, Italy;
abbas.razavykia@polito.it (A.R.); cristiana.delprete@polito.it (C.D.)
* Correspondence: paolo.baldissera@polito.it

Received: 25 August 2019; Accepted: 9 October 2019; Published: 11 October 2019

Abstract: Cryogenic treatment is a supplemental structural and mechanical properties refinement process to conventional heat treatment processes, quenching, and tempering. Cryogenic treatment encourages the improvement of material properties and durability by means of microstructural alteration comprising phase transfer, particle size, and distribution. These effects are almost permanent and irreversible; furthermore, cryogenic treatment is recognized as an eco-friendly, nontoxic, and nonexplosive process. In addition, to encourage the application of sustainable techniques in mechanical and manufacturing engineering and to improve productivity in current competitive markets, cryo-treatment can be considered as a promising process. However, while improvements in the properties of materials after cryogenic treatment are discussed by the majority of reported studies, the correlation between microstructural alteration and mechanical properties are unclear, and sometimes the conducted investigations are contradictory with each other. These contradictions provide different approaches to perform and combine cryogenic treatment with pre-and post-processing. The present literature survey, mainly focused on the last decade, is aimed to address the effects of cryogenic treatment on microstructural alteration and to correlate these changes with mechanical property variations as a consequence of cryo-processing. The conclusion of the current review discusses the development and outlines the trends for the future research in this field.

Keywords: cryogenic treatment; cryo-treatment; mechanical properties; microstructure; cryo-processing

---

1. Introduction

Subzero treatment is a deep stress relieving technology that is associated with cooling components below room temperature. Cryogenic treatment (CT) or cryo-treatment is a subzero heat treatment which is widely used in the production of high precision mechanical parts and components. It provides a large number of applications ranging from industrial components to improvement of musical signal transmission [1,2]. It encourages microstructural refinement and material property improvements such as wear resistance, toughness, fracture resistance, hardness, thermal conductivity, dimensional stability, and chemical degradation [3–8]. Accurate control of process parameters such as cooling rate, soaking temperature and time, cooling fluid, and the whole procedure, promotes the acquisition of superior mechanical properties. In addition, to encourage the application of sustainable techniques in mechanical and manufacturing engineering, and to improve productivity in the current competitive markets, cryo-treatment can be considered as a promising process to improve product service life and reduce production costs in terms of tooling cost and process interruption [9–11]. CT has been classified into three groups: cold treatment, shallow cryogenic treatment (SCT), and deep cryogenic treatment (DCT). Cold treatment is comprised of exposing the ferrous material to be treated in the temperature range of 193–273 K to refine microstructure and achieve better static mechanical properties.

In contrast to heat treating, cold treatment does not require precise control of temperature, and its success depends only on the attainment of the minimum low temperature and is not influenced by lower temperatures [12,13]. In SCT, the samples are placed in a freezer at temperature range of 113–193 K and are then exposed to room temperature. In DCT, the samples are slowly cooled to 77–113 K, held down at this temperature for a certain duration, and are then gradually tempered to room temperature [14,15]. The DCT process consists of three main phases: cooling, soaking time, and warming. Each phase has significant effects on the mechanical properties and microstructure alteration, especially soaking time [16,17]. Generally, DCT is followed by some pre-and post-processing treatments [2,14,18]. While numerous advantages of DCT have been discussed in previous studies and were summarized in a dated review by the authors [14], there is a need for investigating the cryogenic treatment in relation to pre-and post-treatment and their influences on microstructural alteration and mechanical properties. Besides, there are contradictions within the literature due to a lack of understanding of the fundamental metallurgical mechanisms of microstructural alteration and their correlation with the mechanical properties. Therefore, the current review survey is devoted to clarifying the effect of CT on microstructural changes and the correlation between these alterations and the mechanical properties, with a special focus on the knowledge deriving from more recent studies. This review is organized into three main sections: first, all the possible alterations induced by cryo-treatment on the microstructure will be discussed; second, the effects of these microstructural variations on the mechanical properties and their correlation will be addressed; finally, a summary of the conducted experimental investigations on cryo-treatment application in manufacturing engineering will be presented.

2. Effects of CT on Microstructure Alteration

There are imposed stresses whenever a material undergoes any manufacturing process which follows the mechanical component's production. In addition, there are some stresses due to defects in the crystal structure of the material, of which vacancies, dislocations, and stacking faults are the most common [19]. The structural defects are proportional to the level of the imposed stresses when the materials undergo manufacturing processes and cyclic stresses while the materials are under service. Cyclic stresses encourage the defect's migration and their accumulation within the matrix, while the further movement of defects requires more energy. The higher the level of stress, the greater the degree of the defect's migration, which leads to an increase in the interatomic spacing. Consequently, crack nucleation starts and propagates as the distance between the atoms exceeds a critical distance.

For different materials, distinctive alteration scenarios take place in the microstructure during CT [20]. The magnitude of these changes is affected by the treatment process parameters. In most steels, the transformation of retained austenite into martensite [21–24], fine carbide precipitation [25,26], and uniform distribution of secondary carbides [3,27], as well as interstitial carbon atoms segregation, are the common changes observed in the microstructure after CT execution in comparison to the conventional treatments (quenching and tempering) [28]. Applying the appropriate temperature and soaking time causes substantial transformation of the soft austenite with a face centered cubic (FCC) crystal structure into hard martensite with a body centered cubic (BCC) structure, as shown by Figure 1. In addition, metallurgical alteration of martensite at a lower temperature can be obtained [29,30]. These metallurgical changes in martensite are in the form of a reduction in brittleness and an increased resistance against plastic deformation [31]. The formation of fine and very small carbide particles dispersed within the martensite [25,32] provides more interfaces with the base martensite matrix. These interface increments provide more obstacles against dislocation movements and prevents the foreigner particles from penetrating into the matrix, thereby improving the abrasion wear resistance [33,34]. Due to differences in the crystal size of austenite and martensite, there will be imposed stresses where both coexist. CT, and especially DCT, eliminates these stresses by transforming austenite to martensite. Interstitial carbon atom segregation near dislocations is observed during CT, which acts as growing nuclei for the formation of fine carbide particles on subsequent tempering [35,36].

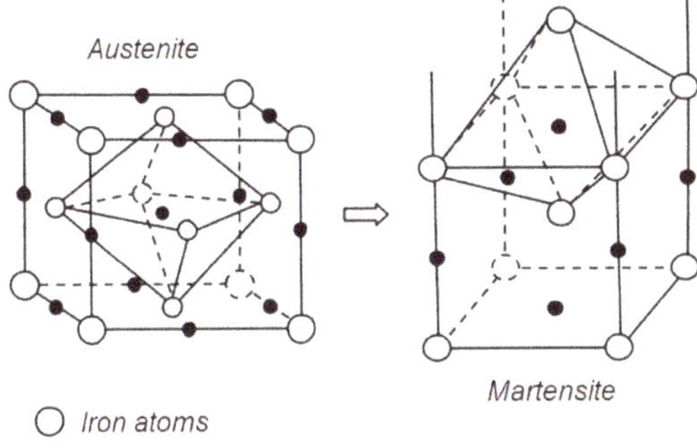

Figure 1. Transformation of soft retained austenite to relatively hard and stable martensite at lower temperature [30].

3. Effects of CT on Microstructure Variation and Hardness

There is a continual demand to improve mechanical properties of industrial components as their complexity grows and functionality requirements increase. Therefore, DCT can be recognized as an alternative process to obtain better mechanical properties with microstructural manipulation. Figure 2 exemplifies a general illustration of the DCT process. DCT improves hardness via two mechanisms, precipitation hardening and martensitic transformation. This section is aimed at providing an insight into the hardening mechanisms contributed by DCT.

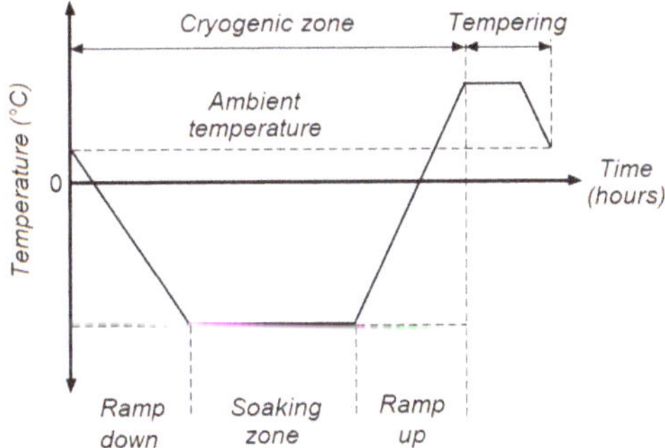

Figure 2. Deep cryogenic treatment (DCT) with single tempering.

A reduction in the level of retained austenite, or even its elimination, can be substantially obtained if DCT is performed efficiently in a vacuum environment. Leskovšek and Ule carried out DCT on AISI M2 high-speed steel to reach a compromise between hardness and fracture toughness [37]. Furthermore, tool shape and dimensional stability have been examined. It was observed that the volume fraction of the retained austenite, and hardness, have a paramount impact on the fracture toughness. DCT promotes a higher rate of retained austenite transition into martensite which improves the hardness, while at the same time, worsens the fracture toughness. Figure 3 shows the effect of DCT on hardness and fracture toughness and their intercorrelation [38]. The combination of DCT and a vacuum environment encourages a higher austenite to martensite transformation and better dimensional stability. In contrast, shape distortion was detected due to the combined effects of transformational and thermal stresses.

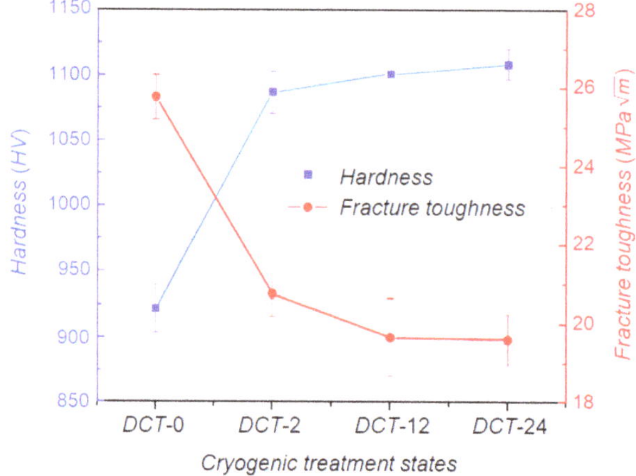

Figure 3. The hardness and fracture toughness of WC-Fe-Ni cemented carbides before and after DCT with different soaking time [38].

Zhirafar et al. conducted an observation to evaluate the alteration of mechanical properties of AISI 4340 after DCT [39]. Higher hardness was detected and consequently, the fatigue limit was improved because of the austenite transformation into martensite. The interaction between hardness and fatigue limit is anticipated to be influenced due to the microstructural changes in steel, considering that the microstructural alteration in steel encourages the achievement of a linear trend between hardness and fatigue limit, even for higher values of hardness [40].

An experimental investigation has been conducted to trace a coherent picture of the effects of DCT on microstructure variation [41]. The complete transformation of austenite to martensite, coupled with a higher volume fraction of fine carbides in the martensite matrix, improved the hardness of En 31 steel as result of DCT and subsequent tempering treatment.

Harish et al. tried to understand the impact of DCT and SCT on En 31 bearing steel microstructural alterations [42]. Higher hardness was obtained through the CT in comparison to conventionally treated samples. Fractography, by means of optical microscopy (OM), provided evidence of equiaxed dimples and flat facets present in the SCT workpiece, and a wide size range of dimples and microcracks in the DCT specimen. Retained austenite and a fine distribution of medium size spheroidized carbide particles have been observed after DCT. Tempered martensite, some spherical carbide particles, and an amount of retained austenite were detected due to SCT execution after tempering. Some needle-like regions were observed in the DCT microstructure prior to tempering, which implies the presence of untampered martensite and the existence of some retained austenite with a lower volume fraction, in comparison to SCT and conventional heat treatment (CHT). Meanwhile, the existence of retained austenite in SCT and DCT structures proved that CT boosts the formation of martensite from retained austenite, but a complete transformation was not observed even after DCT and SCT.

An investigation has been conducted to study the impact of DCT on the microstructure and mechanical property alterations of cold work die steel (Cr8Mo2SiV) [43]. It was observed that the content of precipitation carbides is influenced by the soaking time. DCT results in a martensite and austenite lattice contraction with a homogeneous carbide distribution, and consequently, carbon atoms are forced to diffuse and make a new carbide nucleus. Regarding these alterations, the carbide precipitation content is dependent on the repeating time. Therefore, the volume fraction of the carbide precipitation is affected by both the soaking time and the repeating times. At the end of the process, the hardness of the DCT samples was found to be greater than those obtained by conventionally treated workpieces (quenching and tempering).

HS6-5-2 high speed steel microstructure alteration after DCT has been examined by means of transmission electron microscopy (TEM) and differential scanning calorimetry (DSC) [44]. Qualitatively, it was observed that the martensite obtained after quenching, DCT, and heating up to the ambient temperature was formed by a lamellar-lenticular structure, and internally twinned with a very high density of dislocations. Homogeneous distribution of spherical carbides within the martensite grains is responsible for the hardness improvement.

The effects of SCT and DCT on surface residual stresses, hardness, and impact toughness of 4140 steel have been compared [45]. Higher residual stresses were generated after DCT, in comparison to quenching and further SCT, due to a reduction in the density of lattice defects (dislocations) and thermodynamic instability of the martensite. These effects motivate the movement of carbon and alloying elements toward the defects, and this movement, as mentioned, acts as a basis to form fine carbides after stress relief. It was highlighted that the precipitation of carbides in tempered SCT and DCT samples is responsible for the residual stress relaxation. Meanwhile, the temperature reduction of CT encourages a higher transformation of austenite into martensite and at the same time greater compressive residual stress in the untempered DCT specimens. It is worth mentioning that in contrast to tensile stresses in CHT and SCT workpieces, compressive stresses were observed after DCT.

Mehtedi et al. examined the effects of DCT on treatment response, microstructural alteration, and their correlation with X30 CrMoN 15 1 steel hardness [46]. It was noted that DCT encourages the transformation of the retained austenite to martensite and a homogeneous decoration of the martensitic matrix by refined carbides particles, with a consequent improvement in hardness. It is worth mentioning that, generally, hardness variation is proportional to the austenitizing temperature.

Amini et al. examined the effects of soaking time in liquid nitrogen on the microstructural alteration, carbide distribution, and volume fraction, as well as hardness of 1.2080 tool steel [34]. In all the different soaking times, the elimination of retained austenite, an increment of carbide particles density with homogenous distribution, and a uniform size were observed. Nevertheless, the magnitude of these alterations was affected by the soaking temperature. Hardness improvement was obtained by DCT, but further changes were not noted after a certain holding duration (36 h).

Finite element analysis considering the transformation kinetics has been conducted to simulate DCT and was experimentally validated [47]. Comparison of the experimental data with the numerical method reveals that multiphysical field coupling simulation is an effective and accurate approach to evaluate the cooling behavior of DCT. It was observed that the amount of retained austenite significantly decreased, and as a result, hardness was improved.

Comparative investigation has been conducted to contrast CHT, SCT, and DCT effects on AISI M35 HSS microstructural variations [48]. A higher reduction in the volume fraction of retained austenite was obtained by DCT, followed by SCT. A larger amount of fine precipitations of carbides was observed after DCT. The combination of retained austenite reduction and the distribution of the carbide precipitation resulted in a hardness improvement.

Akhbarizadeh and Javadpour investigated the effect of the as-quenched vacancies on the carbides formation in the microstructure of 1.2080 tool steel after DCT [49]. The effect of the vacancies as potential sites for carbon atoms' jumping during DCT was clarified by using an electric current. It has been noticed that the as-quenched vacancies play a significant role in the carbide formation during DCT by providing appropriate sites for the carbon atoms' jumping, which results in a hardness improvement. The authors concluded that these carbon atoms provide some appropriate places for carbide nucleation during the tempering.

A comprehensive investigation has been conducted to study the microstructural and mechanical properties alteration of ultrafine-grained tungsten carbide–cobalt WC–12Co cemented carbide subjected to DCT. The phase transformation of the binder phase Co and the hardness over a wide range of temperatures was studied using thermal analysis and selective electrolytic corrosion technology as shown in Figure 4. In contrast to previous studies [50], the precipitated tiny second-phase particles could not be observed in the Co binder phase due to magnification limits associated with scanning electron microscopy (SEM). X-ray diffraction (XRD) analysis revealed that DCT promotes transformation of ε-Co, and there is no influence on the crystal structure of tungsten carbide (WC) particles and consequently, this improves the hardness and bending strength of the cemented carbides [51].

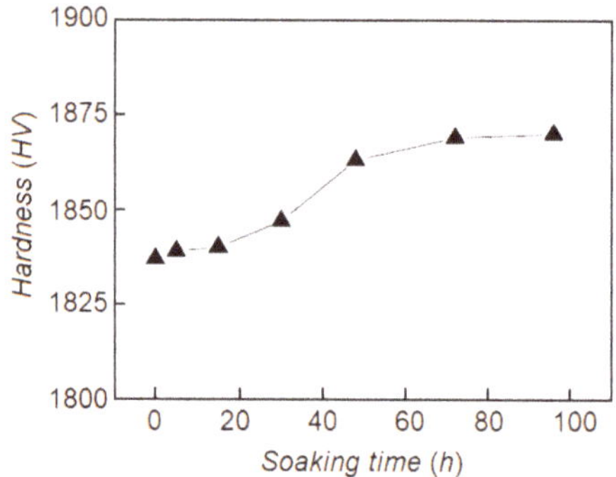

Figure 4. The effect of soaking time on hardness variation of the ultrafine-grained WC–12Co cemented carbide [51].

A comparative investigation between CHT, SCT, and DCT was conducted to evaluate microstructural variation and hardness of AISI 440C bearing steel [52]. A higher rate of retained austenite transformation into martensite was achieved by DCT and a lower volume fraction of martensite was traced by CHT, while DCT samples showed higher hardness.

Xie et al. investigated the influence of DCT on the microstructure and mechanical properties of WC–11Co cemented carbides with different carbon contents [53]. DCT has a nonsignificant effect on phase composition. The amount of η-phase was increased as compared to untreated samples. It was observed that DCT refines the WC grains into triangular prisms with rounded edges, without size alteration, through the spheroidization process. In addition, the phase transformation of the Co phase from α-Co (FCC crystal structure) to ε-Co (HCP crystal structure) was observed after DCT, with reduction of W solubility in the binder (Co). It was highlighted that DCT improved the hardness and bending strength of the alloys, but there was no remarkable impact on the density and cobalt magnetic performance.

Yuan et al. studied the microstructural alteration and mechanical properties of commercially pure zirconium after DCT execution [54]. The DCT reoriented grain is much closer to the (0 0 0 1) basal plane and encouraged a higher grain boundary misorientation and dislocation density. Therefore, the higher fraction of grain boundary misorientation provides more obstacles against the dislocation movement, increasing the material resistance against plastic deformation and improving the hardness. Furthermore, the basal planes are associated with a higher hardness in comparison to the prism planes.

Pérez and Belzunce conducted an observation to examine the influences of DCT on the mechanical properties of H3 tool steel [55]. DCT lessens the retained austenite content in H13 steel, but there is a minimum innate content which cannot be transformed by heat treatment. The H3 steel hardness decreased, as the carbide precipitation and carbon content of the martensite reduced.

Mohan et al. examined the SCT impact on the microstructure and mechanical properties of Al7075-T6 [56]. Static mechanical properties such as hardness, yield strength, and ultimate tensile strength were improved. Precipitation, better distribution of second-phase particles, and higher dislocation density were observed using electron back scattered diffraction (EBSD) after DCT treatment in comparison to untreated specimens, as illustrated in Figure 5. It was highlighted that the hardness and stiffness improvements are the consequence of precipitation hardening and high dislocation density. Fatigue limit has been improved due to striations becoming denser in the cryo-treated alloy.

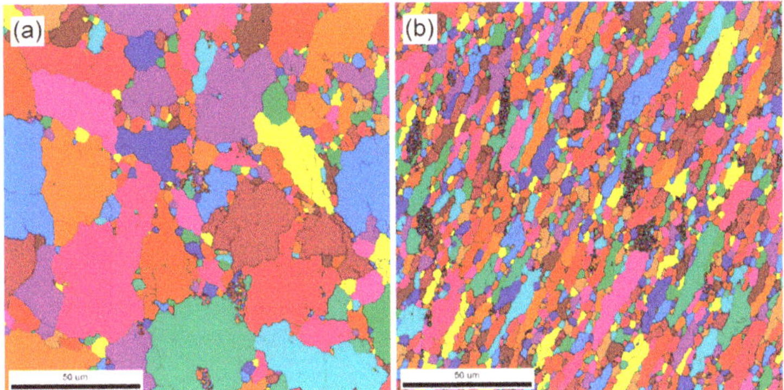

Figure 5. Electron back scattered diffraction (EBSD) micrographs of (**a**) the base and (**b**) the shallow cryogenic treatment (SCT) sample [56].

A set of subzero treatments (SZTs) were conducted on aluminum 2024, to study microstructure, hardness, and tensile and fatigue strength [57]. SZTs were performed at −60 °C (held for 10 h) and at −196 °C (held for 4 h). As the grain size and formed nanoparticles in the microstructure are refined by applying SZTs, elongation, yield strength, hardness, and tensile strength were improved. In contrast, due to microcrack formation in the SZTs, the fatigue limit was reduced under both treatment protocols.

An investigation was designed to examine the effects of the amount of particle refinement and different subzero and ageing processes on the hardness of Al7075 [58]. A smaller atomic distance, refined particle shape, and uniform microstructure were observed after treatments. In addition, the refined microstructure encourages higher friction against dislocation movement, and thus higher hardness.

The current literature survey has a broad consensus about the fact that CT provokes the microstructure alteration and consequent changes in the mechanical properties as summarized by Table 1. In the case of steels, performing a complete treatment process, comprising of austenitization, quenching, cryo-treatment, and tempering, would promote a more refined microstructure and improved mechanical properties. Austenitizing and quenching transform some austenite and primary carbides into martensitic phases. Tempering encourages transition carbides by means of the transformation of supersaturated carbon to form carbides. This contribution relieves microstresses in the martensitic structure and prevents crack nucleation. The effects of CT on steels can be attributed to some important changes in the microstructure that affect the hardness:

- Transformation of retained austenite to martensite can be boosted under vacuum conditions. However, complete transformation cannot be achieved after cryo-treatment as some untempered martensite, characterized by needle-like regions and some volume fraction of retained austenite, has been observed. In this stage, tempering would assist in ameliorating the transformation of retained austenite to martensite and refining the martensitic structure to obtain better hardness;
- Higher volume fraction of fine carbides within the martensite matrix reinforces the martensitic structure and improves hardness;
- Homogeneous distribution of carbides, and a good decoration of the martensite matrix with small size carbide particles, provides more resistance against the dislocation migration within the matrix and plastic deformation;
- Martensite and austenite lattice contractions, along with the uniform distribution of refined carbide particles, encourages carbon atoms diffusion and new carbide nucleation, which results in a higher volume fraction of carbides, especially during tempering, and consequently improves the material hardness;
- Soaking time and temperature, as well as tempering rate, have a prime importance in improving hardness.

Table 1. Summary of literature data devoted to studying the effects of cryo-treatment on microstructure and hardness.

First Author, [#]	Cryogenic Treatment	Rival Treatment	Material	Microstructure Alteration	Outcome
Amini [34]	DCT	N.A.	1.2080 tool steel	Elimination of retained austenite, increment of carbide particles density	Hardness improvement was obtained.
Zhirafar [40]	DCT	N.A.	AISI 4340	Austenite transformation into martensite	Hardness and fatigue limit were improved.
Vimal [41]	DCT followed with tempering	N.A.	En 31 bearing steel	Austenite to martensite transformation coupled with higher volume fraction of fine carbides	Hardness improvement
Harish [42]	SCT and DCT	N.A.	En 31 bearing steel	Distribution of medium size spheroidized carbide particles and exitance of retained austenite even after DCT and SCT	Higher hardness obtained after DCT followed by SCT
Li [43]	DCT	Quenching and tempering	Die steel (Cr8Mo2SiV)	Martensite and austenite lattice contraction and homogeneous carbide distribution	Higher hardness obtained by DCT
Jeleńkowski [44]	Quenching+ DCT+ tempering	N.A.	HS6-5-2	Obtained martensite with lamellar-lenticular structure, and internally twinned, with very high density of dislocations as well as homogeneous distribution of spherical carbides	Hardness improvement
Senthilkumar [45]	SCT and DCT	N.A.	4140 steel	Reduction in lattice defects after DCT and residual stress relief in comparison to quenching+ SCT	Hardness improvement and residual stress releasement
Mehtedi [46]	DCT	N.A.	X30 CrMoN 15 1 steel	transformation of the retained austenite to martensite and homogeneous decoration of martensitic matrix by refined carbides particles	Higher hardness was recorded.
Candane [48]	SCT and DCT	CHT	AISI M35 HSS	Higher reduction in volume fraction of retained austenite was obtained by DCT followed by SCT.	Better hardness obtained by DCT followed by SCT
Akhbarizadeh [49]	DCT+ tempering	N.A.	1.2080 tool steel	Carbon atoms segregation and carbide nucleation	Hardness improvement
SreeramaReddy [50]	DCT	N.A.	WC–12Co cemented carbide	Transformation of ε-Co	Improvement of hardness and bending strength of cemented
Idayan [52]	SCT and DC	CHT	AISI 440C bearing steel	Higher rate of retained austenite transformation into martensite was achieved by DCT followed by SCT	Higher hardness was obtained by DCT.
Xie [53]	DCT	N.A.	WC–11Co cemented carbides	DCT refines WC grains into triangular prism with round edges without size alteration through the spheroidization process	DCT improved hardness and bending strength of the alloys.
Yuan [54]	DCT	N.A.	Pure zirconium	DCT reoriented grain is much closer to (0 0 0 1) basal plane	Increment in material resistance against plastic deformation and improving the hardness
Pérez [55]	DCT	N.A.	H3 tool steel	Reduction in retained austenite content	H3 steel hardness decreased, as carbide precipitation and carbon content of the martensite reduced.
Mohan [56]	DCT	N.A.	Al7075-T6	Precipitation, better distribution of second-phase particles, and higher dislocation density	Hardness and fatigue limit improved.
Nazarian [57]	SCT and DCT	N.A.	Al2024	Grain size and formed nanoparticles in microstructure were refined.	Hardness and fatigue limit were reduced as formation of microcracks.
Taşkesen [58]	DCT	N.A.	Al7075	Smaller atomic distance, refined particle shape and uniform microstructure	DCT encouraged higher friction against dislocation movement and higher hardness.

Cemented carbides after CT undergo some changes within the microstructure, which bring better mechanical properties. Transformation of the binder phase Co from α-Co to ε-Co, an increment of the η-phase volume fraction, and a reshaping of the WC grains without size reduction, are the effects of DCT which improves the bending strength and hardness.

In the case of titanium alloys, DCT imposes an alteration in the volume fraction, size and morphology of the α and β phases and improves hardness.

4. Effects of CT on Microstructure Alteration and Wear Resistance

Enhancement of the wear resistance of loaded components with relative motion is a promising approach to improve efficiency and durability for a wide range of machine elements in tribo-systems. CT would be a reliable option to improve wear resistance of components via material microstructure modifications. Therefore, this section highlights the achievements of previous studies to share the technical synthesis with an emphasis on correlation between microstructural variations and wear behavior.

Thakur et al. compared the impacts of three different post-treatments on tungsten carbide–cobalt inserts' microhardness and microstructural alteration [59]. The effects of controlled cryo-treatment, heating and forced air cooling, and heating and quenching of WC–Co in an oil bath have been investigated. A slight increment of microhardness was obtained by CT and the higher microhardness was measured by the other treatments. A wear resistance improvement after controlled CT was detected due to the densification of the cobalt metal binder which holds the carbide particles firmly.

The effect of austenitizing time in DCT on the wear resistance of D6 tool steel was measured using a pin-on-disk wear test [60]. By increasing the austenitization time from 10 to 50 min, grain size experienced an increment of 122%. It was observed that the diffraction intensities of the martensite increased, while the volume fraction of austenite decreased, which implies that the prior austenite grain size does not affect the retained austenite volume fraction after DCT. After conducting hardness tests, it was concluded that the austenitizing time plays an important role on microstructure homogenization, which encourages higher wear resistance.

Dhokey and Nirbhavne conducted a comparative study to evaluate conventional quenching, tempering, and intermediated CT, and their effects on the wear resistance of D3 tool steel [61]. A huge segregation of carbides of massive size was observed after conventional quenching, but tempering encouraged the reduction in carbide size, and CT resulted in a higher volume fraction of fine carbides and their nucleation during ramp up. Development of fine carbides after CT improves the wear resistance against sliding wear, which could be attributed to the formation of nanosized η-carbides.

A set of treatments were proposed to study the effect of microstructural alteration on wear behavior of D6 tool steel [62]. SCT at −63 °C and DCT at −185 °C were carried out to examine the effects of treatment types, soaking time, and stabilization. Wear tests were conducted applying pin-on-disk wear tests using different loads and sliding speeds. It was observed that CT improved the wear resistance by means of reducing the amount of retained austenite. However, DCT became more efficient in comparison to SCT, as it encouraged a more homogenized carbide distribution and elimination of the retained austenite. The longer the treatment time, the higher the wear resistance and hardness as more retained austenite was transformed into martensite.

Das et al. tried to correlate the soaking time and wear behavior of AISI D2 to find the optimum soaking duration in the range of 0 to 132 h, and a cryogenic temperature of 77 K [63]. Figure 6 demonstrates the effect of soaking temperature on wear resistance of AISI D2 specimens subjected to different loads. The results provide evidence that there is a strong correlation between soaking time and the precipitation behavior of secondary carbides. Wear behavior of the cryo-treated AISI D2 samples is affected by the volume fraction of secondary carbide particles within the matrix. The optimum time of around 36 h for the soaking zone of the considered material was recommended.

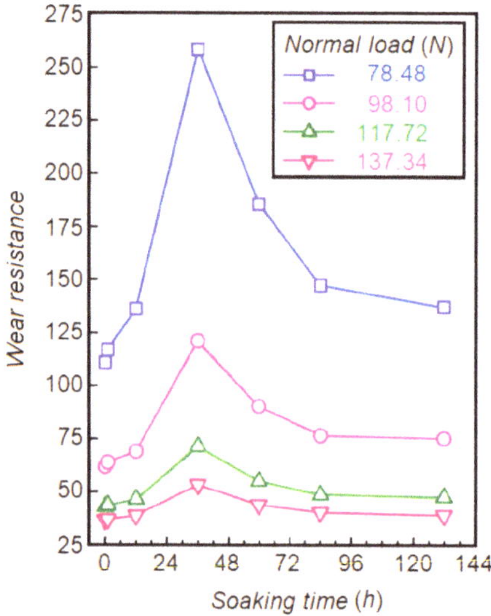

Figure 6. Influence of soaking time on the wear resistance of cryo-treated D2 steel specimens [63].

Straffelini et al. optimized the wear resistance of stamping tools applicable to the automotive industry [64]. The effects of three treatments were examined: coating by thin ceramic film (AlCrN) using physical vapor deposition (PVD), a tool produced by hard metal, and two DCT treated HSS (high-speed steel). DCT improves the wear resistance of the tool as a result of the precipitation of ultrafine carbide particles.

The influence of DCT parameters, soaking time, and temperature on the tribological performance of powder–metallurgy (PM) high-speed steel were investigated by means of abrasive wear resistance and resistance to galling under dry sliding conditions [65]. It was observed that a higher austenitizing temperature leads to a smaller amount of undissolved eutectic carbides of small size. The longer treatment time results in a fine microstructure. It was concluded that the austenitizing temperature is a more significant player in comparison to the soaking time.

Wang et al. studied the effect of DCT on the microstructure and abrasion resistance of a high chromium cast iron [66]. It was observed that the volume fraction of secondary carbides increased after the destabilization treatment combined with DCT. Cryo-treatment resulted in the transformation of the abundant retained austenite into martensite and finer secondary carbide precipitated formation, which are the main contributors to wear resistance improvement. Figure 7 compares the volume fraction of retained austenite after air cooling and cryo-treatment following the destabilization treatment. Therefore, a higher volume fraction of carbide content provides more interface with the matrix, which promotes obstacles against dislocations and enhances the wear resistance.

The effects of DCT parameters, including austenitizing temperature, ramp down, soaking time, ramp up, and tempering temperature, on AISI D2 wear behavior have been examined using the Taguchi method [67]. It was observed that complete transformation of retained austenite to martensite cannot be achieved during ramp down, but the greatest transformation was achieved during soaking time and was affected by soaking temperature.

Das et al. studied the advantages of cooling temperature during subzero treatment on AISI D2 steel wear resistance [12,68]. Wear behavior was evaluated by estimating specific wear rates and conducting detailed characterizations of the worn surfaces, wear debris, and subsurfaces using SEM coupled with X-ray. Figure 8 compares the effects of treatment processes, CHT, and subzero treatments (CT, SCT, and DCT) on the volume fraction of different phases within the matrix. It was observed that the lower the temperature of subzero treatment, the higher the wear resistance. KA reduction in retained austenite volume fraction and the simultaneous increment in the amount of secondary carbide particles at lower temperature improved the wear resistance.

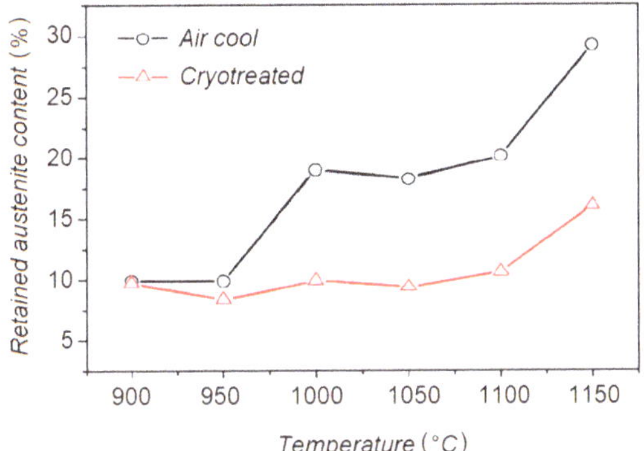

Figure 7. The content of retained austenite at different heat-treated states (heated for 0.5 h) [66].

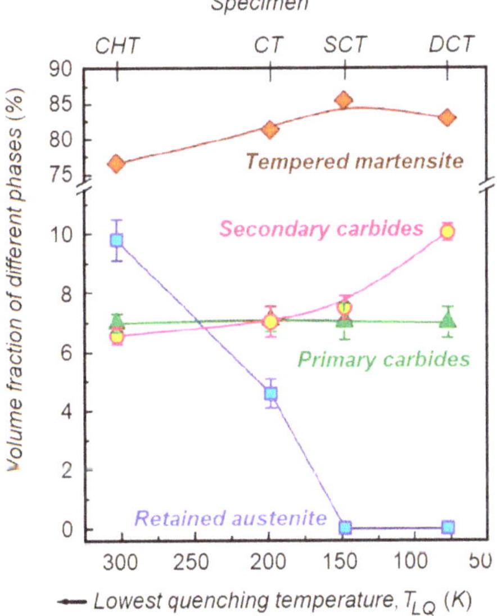

Figure 8. Variation of number of different phases with lowest quenching temperature (TLQ) [68].

Cajner et al. studied the influence of DCT on standard PM S390 MC high speed steels [69]. The effects of DCT on impact and fracture toughness, erosion wear resistance, and microstructure were examined. DCT brought some advantages, such as remarkable improvement of wear resistance due to the presence of η-carbides, however a significant change in toughness was not recorded.

The effects of DCT on microstructure, creep, and wear resistance of AZ91 magnesium alloy have been examined [70]. Alteration of β precipitates distribution, the coexistence of dissolved tiny laminar β particles, and coarse divorced eutectic β phase, were observed as results of DCT. It was highlighted that the creep behavior of the alloy, which is affected by the stability of the near grain boundary microstructure, was improved. Wear resistance was enhanced after DCT as consequences of the internal microstructure stabilization and the formation of a new morphology of β particles.

Sogalad and Udupa conducted an investigation to evaluate the DCT impact on the load bearing capacity of fitted pairs (En8 steel) interference [71]. Different soaking times were applied to the pins sunk in liquid nitrogen and ice. In addition, the bushes were heated to be assembled with the pin without external pressure. Stronger joints were obtained after DCT due to an improvement of the hardness and wear resistance as a consequence of the austenitic transformation to martensite. The improved wear resistance of the pins is attributed to fine carbide formation with a tight lattice structure due to DCT impact. The load carrying capacity of the bearing was improved after cryo-treatment due to a transformation of the retained austenite to freshly formed martensite and the distribution of η-carbides, which fill the microvoids present in the matrix, making it much denser and more uniform.

Akhbarizadeh et al. applied an external magnetic field during DCT to evaluate wear behavior of 1.2080 tool steel [72]. It was observed that DCT reduces the volume fraction of retained austenite and encourages the uniform distribution of carbide particles within the matrix, and consequently, enhances the wear resistance. Surprisingly, the magnetic fields had the reverse effects and reduced the number of carbides, with a nonuniform distribution, which worsened the wear resistance.

Siva et al. investigated the effects of DCT execution on the wear resistance of 100Cr6 bearing steel [73]. Significant improvement of wear resistance after DCT was obtained in comparison to conventionally treated specimens (Figure 9) which was attributed to the alteration of the retained austenite into martensite and the precipitation and distribution of the carbides within the microstructure. DSC revealed that the increment of martensite destabilization by triggering carbon clusterization and carbide precipitation resulted in an improvement of wear resistance and hardness.

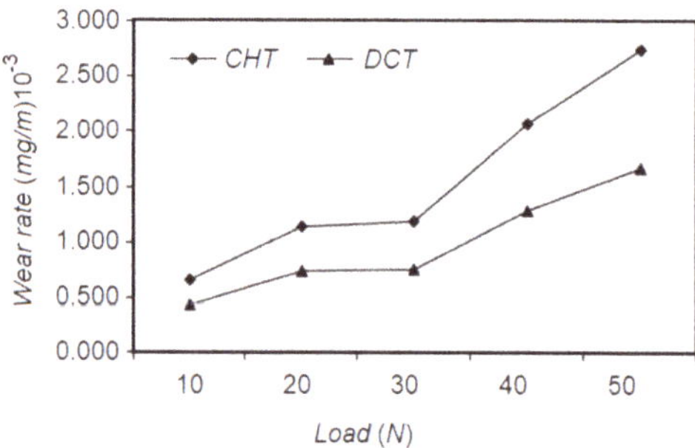

Figure 9. Wear rate of 100Cr6 bearing steel at 5 Hz frequency for the conventional heat treatment (CHT) and DCT [73].

Xu et al. evaluated high performance tool steels' (AISI H13) microstructural alteration prior to and after CT using XRD and synchrotron microdiffraction [74]. The execution of DCT results in diffusion of excess carbon out of the martensite phase and martensitic unit cell shrinkage. Besides these changes, cryogenically treated samples received homogeneous martensite structure. It has been highlighted that a considerable degree of disorder, resulting from rapid cryogenic cooling, encourages microstresses discharge during tempering, which improves wear resistance.

Arockia, Jaswin, and Mohan carried out an observation to optimize the CT with the aim of improving wear resistance, hardness, and tensile strength of E52 valve steel by applying the Grey–Taguchi method [75]. Different combinations of cooling rate, soaking temperature, soaking period, and tempering temperature were considered to execute the DCT. Significant alteration of the retained austenite to martensite and an increment of the fine secondary carbide precipitation were obtained in optimized DCT specimens. Therefore, an improvement of hardness and wear resistance was achieved as a consequence of the aforementioned effects.

Li et al. conducted a set of experimental observations to study the tribological performance of tool steel subjected to DCT [76]. It was highlighted that the segregation of carbon atoms to nearby dislocations, and the interaction produced between themselves and the dislocations, encouraged the formation of carbon clusters. These carbon clusters act as nuclei to form carbide particles during subsequent tempering, and consequently improve wear behavior.

The effects of DCT on the microstructural alteration and wear behavior (abrasion characteristics) of H13 tool steel have been observed [77]. It was highlighted that the lower the temperature, the higher the amount of transformed retained austenite to martensite, resulting in smaller and more uniform distributions of martensite laths. Furthermore, DCT promotes the precipitation of more homogeneous and very fine carbide particles. The joint effects of these microstructural variations improved the wear properties of the H13 tool steel.

Senthilkumar and Rajendran examined the effects of DCT on the wear resistance of En19 steel [78]. Three different types of heat treatment were performed. The influences of DCT, SCT, and CHT were investigated through dry sliding wear testing. X-ray observation revealed that the precipitation of fine carbides and the transformation of retained austenite into martensite, during DCT and SCT, improved the wear resistance in comparison to CHT. It was highlighted that a higher wear resistance and lower friction coefficient have been obtained by DCT.

Jaswin et al. evaluated the effects of SCT and DCT on the microstructure and wear resistance of X45Cr9Si3 and X53Cr22Mn9Ni4N valve steels [79]. The alterations were compared to conventionally heat-treated specimens. SEM analysis reveals that the full elimination of the retained austenite is not achievable even after performing SCT and DCT, but a reduction in the retained austenite was observed as a consequence of CTs. It was highlighted that the formation of fine carbides dispersed in the tempered martensite structure is the main contributor to improved wear resistance, followed by the retained austenite transformation.

Li et al. carried out a set of experimental trials to evaluate DCT effects on internal friction behaviors in the process of tempering [80]. Wear resistance improvement was observed due to greater carbide precipitation after DCT by means of applying internal friction. This is because as the carbon atoms move towards the dislocations, strong interaction is generated between interstitial carbon atoms and among them with a time dependent strain field of dislocations. The generation of carbon atom clusters adjacent to the dislocations under DCT, inspires carbide formation which subsequently results in wear resistance improvement.

Amini et al. investigated the effects of stabilization, tempering, and DCT on the mechanical properties of 80CrMo12 5 tool steel [81]. Elimination of the retained austenite and a higher amount and finer distribution of carbide, as a consequence of DCT, dramatically improved the wear resistance. Ultimate tensile strength increased and a reduction in specimen toughness were also observed. It was concluded that DCT treatment should be done immediately after quenching to obtain the highest wear resistance and hardness.

Gill et al. conducted a set of trials to examine the effects of DCT on the mechanical and microstructural alteration of AISI M2 HSS [82]. It was proved that complete transformation of austenite into martensite cannot be achieved by DCT, however some advantages, such as higher precipitation of small carbides, an increment of their volume fraction, and uniform distribution of the carbides, have been obtained after DCT followed by SCT, as shown in Figure 10. Wear resistance improvement was observed due to the aforementioned DCT and SCT effects.

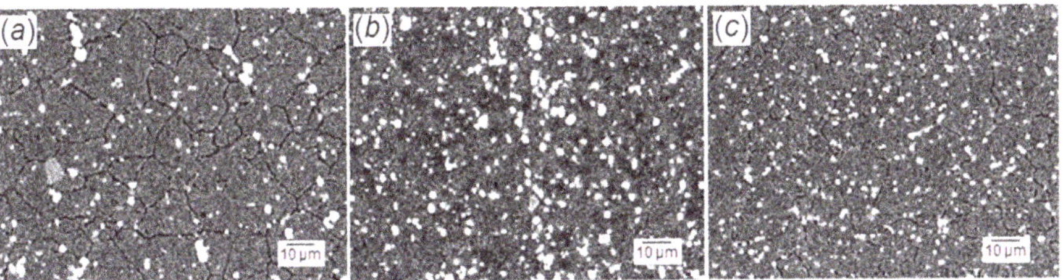

Figure 10. SEM image used for quantitative determination of the size and volume fraction of carbides for (**a**) CHT, (**b**) SCT, and (**c**) DCT [82].

Sri Siva et al. applied the Taguchi method and Gray relational analysis to optimize the DCT processes to be executed on 100Cr6 bearing steel [83]. Dimensional stability, wear resistance, and hardness were selected as the indicators to examine the effect of cooling rate, soaking temperature, and duration, as well as tempering temperature, to be proposed for optimization. It was highlighted that the precipitation of the fine carbides and the transformation of the retained austenite to martensite are the main contributors to the improvement of hardness, wear resistance, and dimensional stability.

Isothermal martensitic transformation at temperatures below $-100\,^{\circ}\text{C}$ during DCT was studied as a physical mechanism that has a significant contribution to mechanical properties improvement [84]. Steel X153CrMoV12 was selected on which to conduct the observations. Findings revealed that the joint effects of low temperature martensitic transformation and plastic deformation play a remarkable role in the subsequent carbide precipitation, which improved the mechanical properties such as abrasive wear resistance and hardness. Low-temperature isothermal martensitic transformation, as a result of unbalanced volumes of the martensite and retained austenite, leads to plastic deformation which encourages the refining of primary and secondary carbides and consequently, improves hardness.

A study was designed to determine the effects of influential variables on the correlation between treatment parameters and mechanical properties of AISI D2 steel subjected to distinct categories of SZTs viz. cold, SCT, and DCT [85]. The effects of the treatment process were examined by means of the number of different phases and stereological parameters of the secondary carbide (SC) particles. It was highlighted that a reduction in the retained austenite, and at the same time, an increment of SCs by subzero treatment, resulted in an improvement of bulk hardness and microhardness with a minor fracture toughness loss, and a remarkable improvement in wear resistance.

The effects of DCT on D2 tool steel wear resistance have been studied by means of pin-on-disk tests using steel and tungsten carbide pins [86]. Microstructural alteration and wear mechanism were evaluated using SEM and XRD. Wear behavior was improved due to the formation of fine carbides with the size varying from micron to nanoscale. In addition, a higher amount of carbide content and homogeneous carbide distribution were observed after DCT treatment. Adhesive wear was the predominant mechanism under different observational set-ups (load, sliding speed, and pin material).

Das and Ray studied the mechanism of wear resistance improvement by applying DCT [87]. It was highlighted that DCT eliminates retained austenite and encourages the formation of secondary carbides and consequently enhanced the wear resistance of AISI D2 steel.

Podgornik et al. reported that DCT improved the abrasive wear resistance and better galling properties of powder–metallurgy (PM) high-speed steel, as a result of a finer needle-like martensitic microstructure formation which provides a lower average friction coefficient, while applying longer soaking time [88].

The effects of different soaking and tempering times during DCT of 45WCrV7 tool steel have been studied [89]. Maximum hardness was obtained by increasing either the soaking time or tempering time, as these two encouraged the higher transformation rate of retained austenite into martensite, in addition to a larger volume fraction of carbides with a homogenous distribution and uniform size. It was proved by another investigation that the joint effect of soaking and tempering times is the most influential parameter to enhance hardness and consequently, wear resistance [90].

An AZ91 magnesium alloy was subjected to DCT and the microstructural alteration was examined using OM and SEM [91]. Hardness and wear tests were performed to evaluate the effects of DCT and microstructural changes as a consequence of the treatment. DCT improves hardness and wear resistance of AZ91 by structure contraction, in which aluminum atoms in the β phase jump adjacent to the defects as new Mg17Al12 precipitates during ageing. As the expansion coefficient of α and β phase are different at a lower temperature, this difference is associated with different shrinkage values. As consequence of these different shrinkage values, dislocations and microvoids are generated on the boundary of the α and β phase that provide appropriate places for aluminum atoms to jump during DCT.

An investigation has been conducted to determine the impact of DCT on AISI 52100 bearing steel wear resistance [92]. Significant microstructural alteration and consequently higher wear resistance were obtained by DCT execution. Uniform carbide distribution and homogeneous particle size along the formation of new small size carbides were obtained after DCT, which resulted in higher wear resistance. Better hardness was achieved due to the redistribution of chromium carbides and the austenite to martensite transformation. It was highlighted that the soaking time is the most important process parameter to obtain a uniform carbide distribution, fine and homogeneous particle size, and better mechanical properties. Figure 11 provides evidence that until a certain duration of soaking time (12 to 36 h), positive effects on carbide particle size and distribution can be observed, but surprisingly, the opposite effects can be seen for a longer holding time.

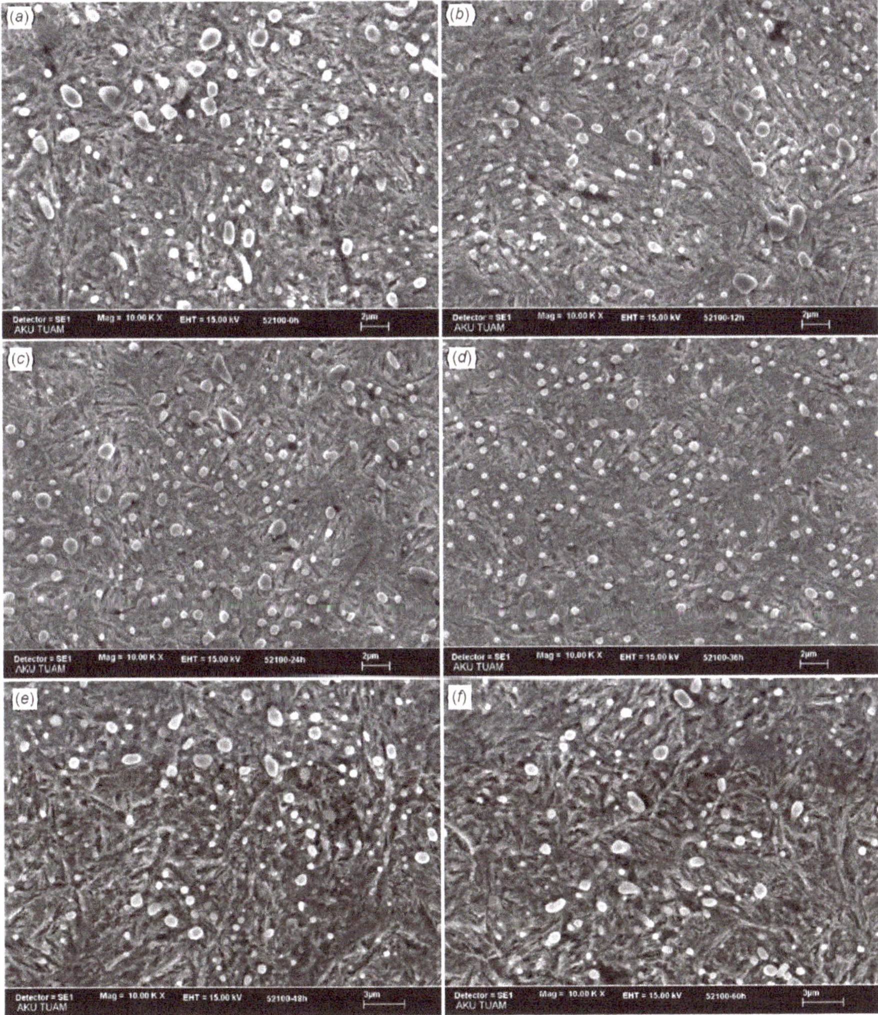

Figure 11. SEM microstructures of (**a**) CHT, (**b**) DCT-12 h, (**c**) DCT-24 h, (**d**) DCT-36 h, (**e**) DCT-48 h, and (**f**) DCT-60 h samples [92].

The effects of different treatments on AISI D3, as-received, CH treated without tempering, and DC treated without tempering were evaluated [93]. Uniform distribution of primary and secondary chromium carbides with smaller sizes were promoted by CHT and DCT. The highest hardness was obtained by DCT followed by CHT and consequently better tribological performance was achieved by DCT.

Microstructure and wear properties of dies are essential parameters that affect the product quality and production costs. Therefore, the effects of treatment processes on the microstructure and wear resistance of AISI H13 hot-worked tool steel as the most common material in die production, have been examined [94]. Different treatment processes, conventional, DCT, and DCT plus tempering (DCTT), were considered and executed. It was noted that DCT encourages a homogeneous distribution of fine carbide particles in the martensite matrix and improves the wear resistance, but better wear resistance is obtained by DCT due to the formation of finer carbides with a more uniform distribution. In addition, tempering after DCT resulted in the precipitation of secondary carbides and improved their coherence with the matrix.

Li et al. performed experimental trials to study the effects of DCT on high-vanadium alloy steel microstructure and mechanical properties [95]. Enormous amounts of small secondary carbide precipitation and microcracks at the carbide and matrix interfaces were observed after execution of DCT. It was highlighted that the volume fraction of secondary carbides after DCT was found to be three to five times larger than those obtained by CT. This increment in the volume fraction of secondary carbides was found to be affected by the number of DCT cycles. In addition, DCT encourages the finer carbide particle size with a homogenous distribution. The DC treated sample formed smaller secondary spherical carbides. In contrast, as hardness decreases, toughness increases after DCT. A slight improvement of abrasive wear resistance was obtained after DCT execution, due to a larger amount of finer secondary carbide precipitation, which is affected by soaking time and the number of cycles.

An investigation was aimed at studying the effects of DCT on fracture toughness, wear resistance, and load-carrying capacity of cold-worked tool steel [96]. It was concluded that an improvement in the mechanical properties was as a consequence of finer, needle-like martensite formation and retained austenite transformation, combined with the plastic deformation of primary martensite. In addition, any alteration of fracture toughness to working hardness influences the wear resistance of cold-worked tool steel.

The effects of DCT and the correlation between microstructural alteration and wear behavior of WC–Fe–Ni cemented carbides has been examined and discussed [38]. Selective electrolytic corrosion technique was employed to study the phase composition and quantitative analysis of DCT. Wear resistance improvement was observed after execution of DCT due to the martensite phase transformation from γ-(Fe,Ni) to α-(Fe,Ni) and can be attributed to precipitation of W particles in the binder phase. It was highlighted that soaking time is the most important contributor to transformation of the binder phase from γ to α.

Li et al. applied DoE (Design of Experiment) orthogonal design to evaluate the effects of cryogenic-ageing circular treatment (CACT) on articles reinforced in-situ aluminum matrix composites [97]. Different combinations of cooling rate, soaking time and circular index as process variables were set. Al–Zn–Mg–Cu wrought aluminum reinforced by alumina particles, was selected as the host material to execute CACT. SEM analysis revealed that increasing the precipitation of Si phases improved the load carrying capacity of the metal matrix and consequently, the wear resistance. Hardness was reduced as a consequence of a reduction in the unstable but hard needle-like η' (Zn_2Mg) phase and an increment of the stable lamellar h (Zn_2Mg) phase with low hardness.

In many mechanical applications, especially in automotive and aerospace engineering, as the components are reciprocating or rotating at relative speed to each other, wear resistance improvement is of paramount importance. The benefits of CT and its effects on the enhancement of wear resistance have been investigated and discussed in the analyzed literature as summarized by Table 2. However, the mechanisms responsible for the wear resistance improvement by DCT and SCT, in terms of microstructural alteration, are yet to be clarified. Therefore, the impacts of microstructural variation after CT on the wear resistance of steels can be summarized as below to correlate the wear behavior with microstructural alteration:

- Transformation of retained austenite to martensite, which is counted as a contributor to improve microhardness and enhancement of wear resistance to comply with it;
- Higher volume fraction of fine, and a homogeneous distribution of, carbide particles within the matrix resist against the dislocation movements and plastic deformation;
- Segregation of carbides and the formation of η-carbides improves the resistance against sliding wear as they reinforce the martensitic structure;
- Secondary carbide precipitation, especially during tempering;
- DCT is more efficient to enhance the wear resistance in comparison to SCT as it encourages a more uniform distribution of fine carbides and the elimination of retained austenite; soaking time and soaking temperature are the most effective process parameters.

In the case of cemented carbides, DCT enlarges grain size as a consequence of carbide grain contiguity and the dominant effect of the carbide phase, improving thermal conductivity and consequently, wear resistance. In addition, distribution of η-phase carbide within the structure empowers the bond between the carbide and binder, which brings higher wear resistance.

Table 2. Summary of the literature data devoted to studying the effects of cryo-treatment on microstructure and wear resistance.

First Author, [#]	Cryogenic Treatment	Rival Treatment	Material	Microstructure Alteration	Outcome
Thakur [59]	DCT	N.A.	Tungsten carbide–cobalt	Densification of the cobalt metal binder	Wear resistance has been improved.
Akhbarizadeh [60,62,72]	SCT and DCT	N.A.	D6 tool steel and 1.2080 tool steel	Higher volume of homogenized carbide distribution and elimination of the retained austenite	DCT homogenizes microstructure which encourages higher wear resistance
Dhokey [61]	DCT	Quenching and tempering	D3 tool steel	Higher volume fraction of fine carbides and their nucleation during ramp up	Wear resistance improvement.
Das [63,68,83,87]	SCT and DCT	DHT	AISI D2	Reduction in retained austenite and higher volume fraction of secondary carbides	Wear behavior is proportional to secondary carbides morphology. DCT is most effective treatment.
Straffelini [64]	DCT	AlCrN PVD coating and	Stamping tools	Precipitation of ultrafine carbide particles	Wear behavior was improved.
Podgornik [65,88,97]	DCT	N.A.	Powder–metallurgic (PM) high-speed steel and Cold-work tool steel	Homogenous microstructure, finer needles like martensite formation and retained austenite elimination	Abrasive wear resistance has been enhanced.
Wang [66]	DCT	N.A.	High chromium cast iron	Transformation of abundant retained austenite into martensite and finer secondary carbide precipitation	Wear resistance improvement was recorded.
Oppenkowski [67]	DCT	N.A.	AISI D2	Transformation of retained austenite into martensite	Transformation of retained austenite to martensite is affected by soaking time and temperature.
Cajner [69]	DCT	N.A.	PM S390 MC high speed steels	Fine η-carbides formed within the matrix.	Wear behavior was improved.

Table 2. *Cont.*

First Author, [#]	Cryogenic Treatment	Rival Treatment	Material	Microstructure Alteration	Outcome
Asl [70]	DCT	N.A.	Magnesium alloy AZ91	Alteration of β precipitates distribution, coexistence of dissolved tiny laminar β particles and coarse divorced eutectic β phase	Wear resistance was enhanced.
Sogalad [71]	DCT	N.A.	En8 steel	Transformation of retained austenite to martensite and distribution of fine η-carbides	Wear resistance and load carrying capacity were improved.
Siva [73]	DCT	CHT	100Cr6 bearing steel	Martensite destabilization by triggering carbon clusterization and carbide precipitation	Wear resistance and hardness were improved.
Xu [74]	DCT	N.A.	AISI H13	Homogeneous martensite	Wear resistance was enhanced.
Arockia Jaswin [75]	DCT	N.A.	E52 valve steel	Transformation of retained austenite to martensite and increase in the amount of fine secondary carbide precipitation	Wear behavior was ameliorated.
Li [76]	DCT + tempering	N.A.	Tool steel	Precipitation of carbide particle and carbon segregation nearby dislocation and carbon cluster formation	Improvement of wear resistance was highlighted.
Koneshlou [77]	DCT	N.A.	H13 tool steel	Uniform distribution of martensite laths and transformation of retained austenite to martensite	Wear properties has been improved.
Senthilkumar [78]	SCT and DCT	CHT	En19 steel	Precipitation of fine carbides and transformation of retained austenite into martensite	Higher wear resistance was obtained after DCT followed by SCT.
Jaswin [79]	SCT and DCT	CHT	X53Cr22Mn9Ni4N valve steels	Elimination of the retained austenite and formation of fine carbides	Wear resistance was improved.
Li [80]	DCT	N.A.	Cold work die steel	Carbon atoms segregation and generation of strong interaction with dislocations	Wear resistance improvement
Amini [81,86]	DCT	N.A.	80CrMo12 5 tool steel, D2 tool steel	Elimination of retained austenite, higher amount, and fine distribution of carbide	Wear resistance and ultimate tensile strength were improved.
Gill [82]	SCT and DCT	N.A.	AISI M2 HSS	Transformation of austenite into martensite and higher precipitation of small carbides	Wear behavior was ameliorated.
Sri Siva [83]	DCT	N.A.	100Cr6 bearing steel	Precipitation of the fine carbides and transformation of the retained austenite to martensite	Wear resistance, hardness, and dimensional stability were improved.
Gavriljuk [84]	DCT	N.A.	Steel X153CrMoV12	Refining of primary and secondary carbides	Abrasive wear resistance and hardness were enhanced.
Amini [91]	DCT	N.A.	AZ91 magnesium alloy	Aluminum atoms in the β phase jump adjacent to the defects	Wear and hardness were improved.
Gunes [92]	DCT	N.A.	AISI 52100	Uniform carbide distribution, refinement of particle size and redistribution of chromium carbides	Higher wear resistance was obtained.
Khun [93]	DCT	CHT	AISI D3	Uniform distribution of primary and secondary chromium carbides	Higher wear behavior and hardness were recorded.

Table 2. *Cont.*

First Author, [#]	Cryogenic Treatment	Rival Treatment	Material	Microstructure Alteration	Outcome
Çiçek [94]	DCT and DCTT	N.A.	AISI H13	Finer carbides size and distribution	Wear resistance was improved. Tempering after DCT resulted in precipitation of secondary carbides.
Li [95]	DCT	N.A.	High-vanadium alloy steel	Finer carbide particles size with homogenous distribution	Higher abrasive wear resistance was obtained.
Li [97]	DCT	N.A.	Al–Zn–Mg–Cu wrought aluminum	Precipitation of Si phases and reduction in unstable but hard needle-like η' (Zn2Mg) phase	Wear performance and hardness were improved.

5. Effects of DCT on Microstructure and Mechanical Properties

In this section, the effect of CT on fatigue strength, tensile toughness, thermal conductivity, and durability, as well as residual stresses relief, will be addressed and discussed.

Bensely et al. conducted an experimental investigation to gain insight into fatigue and fracture behavior of carburized EN 353 steel subjected to CT [98]. Rotating bending fatigue tests were carried out to study the influences of SCT and DCT. It was highlighted that due to the presence of higher retained austenite and fine carbides, SCT treated specimens showed better fatigue life in contrast to DCT and conventionally treated specimens. A 71% improvement in fatigue life was observed via SCT and DCT.

Baldissera and Delprete studied the effect of DCT on tensile properties, fatigue life, and corrosion resistance of AISI 302 stainless steel in both hardened and solubilized states [99,100]. Although no significant effects on tensile strength were observed, changes in hardness and in elastic modulus were detected as far as significant increases in fatigue life for the solubilized material only. The fatigue fracture surface analysis pointed out that small secondary cracks for the treated material have been observed that could have acted as energy absorber with respect to the cyclic loading. The absence of significant effects from DCT applied to hardened AISI 302 was confirmed following an experimental investigation by Baldissera et al., where the same material was additionally subjected to CrN coating through Physical Vapour Deposition (PVD) [101].

The same authors carried out experimental investigations to assess the effects of DCT on 18NiCrMo5 carburized steel [102,103]. Two levels of temperature and soaking time were considered. In addition, the influence of the sequence among case hardening, tempering, and DCT was studied. Performing DCT before tempering produced a significant increase in hardness (+2.4 HRC), while pre-tempering DCT resulted in a remarkable enhancement of UTS (Ultimate tensile strength) (+11%). The most interesting result was observed in terms of scatter reduction of fatigue data, leading to a drastic increase in the fatigue limit at higher reliability levels (+25% considering a failure probability of 0.3%). In a subsequent study, the authors applied the Tanaka–Mura model for fatigue crack nucleation to the above experimental results through Maximum Likelihood Estimation (MLE), obtaining further insight on the potential causes of the observed improvements [104]. The conclusion was to address further investigations towards the effect of DCT on the specific fracture energy and, in agreement with most of the literature, on the dimensional and statistical characterization of subgrain precipitate fields.

Bouzada et al. evaluated microstructural alterations due to DCT of a heat-treated 7075 aluminum alloy [105]. Higher compressive residual stresses and slight changes in static mechanical properties (yield strength, tensile strength, and hardness) were observed after treatment. Accretion of submicrometric particles near and at the grain boundaries was detected. The combined effect of these microstructural changes improved the stress corrosion cracking and durability of 7075 aluminum.

The joint properties of 4 mm-thick 2219-T87 alloy produced by tungsten inert gas welding (TIG) and friction stir welding (FSW) at room temperature and DCT temperature ($-196\ °C$) were examined. FSW presents better welding properties due to the preservation of the working structures and homogenous chemical compositions. A uniform and coherent microstructure obtained by DCT in the form of grain-boundary strengthening, substructure strengthening, and ageing precipitation strengthening is the main reason for the mechanical properties alteration [106].

The effect of microstructural alteration after DCT execution on tensile toughness in medium carbon-low alloy tool steel has been examined [107]. It was highlighted that the amount of secondary carbides is proportional to the soaking time. As soaking time increases, there was a constant increment of secondary carbides until a certain duration because after this point secondary carbide density is reduced. Tensile toughness was increased by simultaneously extending both the tempering duration and soaking time.

The combined effects of postweld heat treatment and low-temperature ageing treatment with and without DCT on friction-stir-welded joints of 2024-T351 aluminum alloys were examined [29]. It was observed that joint elongation increases with an improvement in tensile strength. Pre-DCT promotes the redissolution or dispersed precipitation of the unstable phases in the as-welded joints.

The effects of DCT on the fatigue behavior of 1.2542 tool steel have been examined [108]. It has been highlighted that nonsignificant changes in fatigue life were observed and both specimens were free of retained austenite. However, specimens that experienced DCT contained more secondary carbides with uniform distribution.

The effects of DCT on mechanical properties of induction hardened En 8 steel were measured, with the treatment carried out at $-196\,^{\circ}\text{C}$ for 24 h [109]. A remarkable improvement of the ultimate tensile strength was observed as DCT increased the compressive residual stress of steel. SEM analysis provided evidence that the martensite structure was formed after DCT.

Araghchi et al. applied a combination of prior treatments, CT, and ageing (reheating by oil at $180\,^{\circ}\text{C}$) on 2024 aluminum [110]. Application of this treatment cycle resulted in a reduction of residual stress by 250%. The concerned treatment encouraged the formation of large needle-like S′ precipitates with different orientation, which improved matrix hardness and reduced the residual stresses. It reported that S′ phase preferentially precipitated on dislocations.

An investigation was conducted to examine the effects of DCT, prior and after ageing, on microstructural evolution, microhardness, and tensile property variations of TB8 metastable β titanium alloy [111]. It was observed that conventional solution treatment followed by DCT resulted in a higher degree of super-cooling and internal stress through lattice shrinkage, which fosters the formation of needle-like α phase in the martensite matrix and furnishes a higher fraction of α phase during ageing treatment. The aforementioned microstructure alterations, as well as refinement and homogenization of lath-like α precipitates, improved the microhardness and tensile strength while reducing the elongation.

Subzero treatment was carried out on EN 1.4418 steel to evaluate the sensitivity of reverted austenite to temperature [112]. It was highlighted that impact toughness is a time dependent variable of temperature and it is directly affected by the volume fraction of austenite. Reverted austenite appears to be stable at room temperature and at elevated temperature (immersing in boiling N2), but in moderate temperature it can be partially transformed.

In general, according to the above literature survey, as summarized by Table 3, some important microstructural alterations due to CT and their effects on the mechanical properties are correlated with each other as follows:

- Residual stresses can be imposed by external or internal sources to the materials. Residual stresses due to external sources can be introduced and developed during manufacturing processes such as nonuniform plastic deformation under cold working, shot peening, hammering, grinding, or by welding due to thermal shocks. Some residual stresses are due to defects in the crystal structure of the materials, with the most common defects being in the form of vacancies, dislocations, and stacking faults. Residual stresses would cause creep failure, fatigue, and stress corrosion cracking in sensitive materials. The higher transformation of retained austenite to martensite at lower temperatures induces compressive residual stress, which is beneficial for wear resistance improvement and fatigue behavior. Precipitation of the carbides in DCT and SCT followed by tempering is the main contributor to residual stresses reduction;

- Fatigue life improvement due to CT is attributed to distribution of fine carbide particles and reduction of residual stresses. Moreover, the subgrain carbide distribution can play an important role by affecting the statistical dispersion of the fatigue behavior, with a significant impact at high reliability levels, which is the most interesting in many structural component applications;
- Fracture toughness after DCT reduced the transformation of retained austenite to martensite. DCT improves hardness and reduces fracture toughness. Therefore, it is essential to control the fracture toughness to avoid the risk of microcracking, especially in the case where mechanical components are sliding under the influence of a load in tribo-systems;
- Formation of martensite and the elimination of retained austenite, along with the precipitation of fine carbide, are the main contributors to the enhancement of ultimate tensile and yield strength.

Table 3. Summary of literature data devoted to study the effects of cryo-treatment on microstructure and mechanical properties.

First Author, [#]	Cryogenic Treatment	Rival Treatment	Material	Microstructure Alteration	Outcome
Bensely [98]	SCT and DCT	N.A.	EN 353	Higher volume fraction of retained austenite and fine carbides	SCT treated specimens show better fatigue life in contrast to DCT and CHT.
Baldissera [99–101]	DCT	CrN coating through PVD	AISI 302	Formation of microsecondary cracks on surface	Tensile strength and fatigue life significantly improved.
Baldissera [102–104]	DCT	N.A.	18NiCrMo5 carburized steel	N.A.	Hardness (+2.4 HRC) and UTS (+11%) were improved.
Bouzada [105]	DCT	N.A.	7075 aluminum alloy	Accretion of submicrometric particles near and at the grain boundaries	Yield strength, tensile strength and hardness were ameliorated
Lei [106]	DCT	N.A.	2219-T87	Uniform grain-boundary strengthening, substructure strengthening, and aging precipitation strengthening	Better weldability obtained.
Vahdat [107]	DCT	N.A.	Medium carbon-low alloy tool steel	Formation of secondary carbides	Tensile toughness increased.
Niaki [108]	DCT	N.A.	1.2542 tool steel	Elimination of retained austenite and uniform distribution of secondary carbides	Nonsignificant changes in fatigue life were recorded.
Senthilkumar [109]	DCT	N.A.	En 8 steel	Uniform martensite structure	Significant improvement of ultimate tensile strength was traced.
Araghchi [110]	DCT	N.A.	2024 aluminum	Formation of large needle-like S′ precipitates with different orientation	Reduction of residual stress by 250% and hardness increase have been reported.
Gu [111]	DCT	N.A.	TB8 metastable β titanium alloy	High volume fraction of needle-like α phase in martensite matrix	Microhardness and tensile strength were improved.
Nießen [112]	DCT	N.A.	EN 1.4418 steel	Stabilization of reverted austenite	Impact toughness is a time dependent variable on temperature, and it is directly affected by volume fraction of austenite.

6. Cryogenic Application in Manufacturing Engineering

Cryogenic treatment induces significant effects on engineering components and improves the mechanical properties through microstructural alteration. A major portion of manufacturing cost is contributed by tooling cost involving tool replacement and resharpening, material scrap, and production line interruption [113]. Improving cutting tool durability and workpiece machinability provides evidence to consider cryogenic treatment as a potential way to take a step toward sustainable manufacturing and clean production. In addition, in such applications where wear resistance is of paramount importance, such as tribo-systems and transportation applications, cryogenic treatment can improve the system reliability. In this context, to improve productivity, it is essential to develop tools with the capability to withstand higher cutting speeds and feed rates during machining. To achieve this goal, tool performance and machinability improvement of workpiece materials are of prime importance, especially in the case of hard to cut materials, as they are most in demand to operate under severe conditions [114,115]. Therefore, this section is devoted to addressing the application of cryogenic treatment in machining processes.

Yong et al. investigated the effects of cryogenic treatment on the cutting performance of tungsten carbide (G10E) during milling of ASSAB 760 medium carbon steel with and without the presence of coolant. The cryo-treated tool showed a better performance especially during wet milling, as DCT improves the thermal conductivity of the tool and promotes higher heat transfer at the tool/chip interface temperatures [116].

The effects of CT on wear resistance and tool life of M2 HSS drills during high speed dry drilling of normalized CK40 carbon were measured. The performance of three different tools, untreated, cryo-treated, and cryo-treated followed by tempering were examined. The findings revealed that cryo-treated drills showed better wear resistance to diffusion wear due to the formation of a higher volume fraction of fine and homogeneous carbide particles in the matrix as demonstrated by Figure 12 [117].

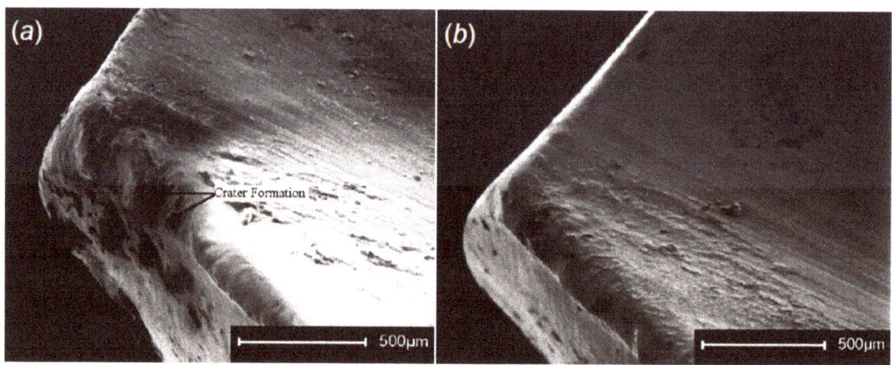

Figure 12. SEM images of drills after drilling of 30 holes: (**a**) nontreated drill and (**b**) cryo-treated with tempering drill [117].

Tool wear and power consumption were considered as indicators to evaluate coated carbide inserts subjected to CT during turning of AISI/SAE 80-55-06 SG iron. Cryo-treated inserts provided better performance due to a reduction of the η-phase (carbide) effect on the hardness and wear resistance by CT, as it promotes a uniform distribution of refined carbide particles and their denser and more coherent bonding with the matrix. Figure 13 shows flank and crater wear of cryo-treated (a and c) and untreated (b and d) cutting tools at a cutting speed of 250 m/min, feed rate of 0.1 mm/rev, and depth of cut 0.8 mm [118].

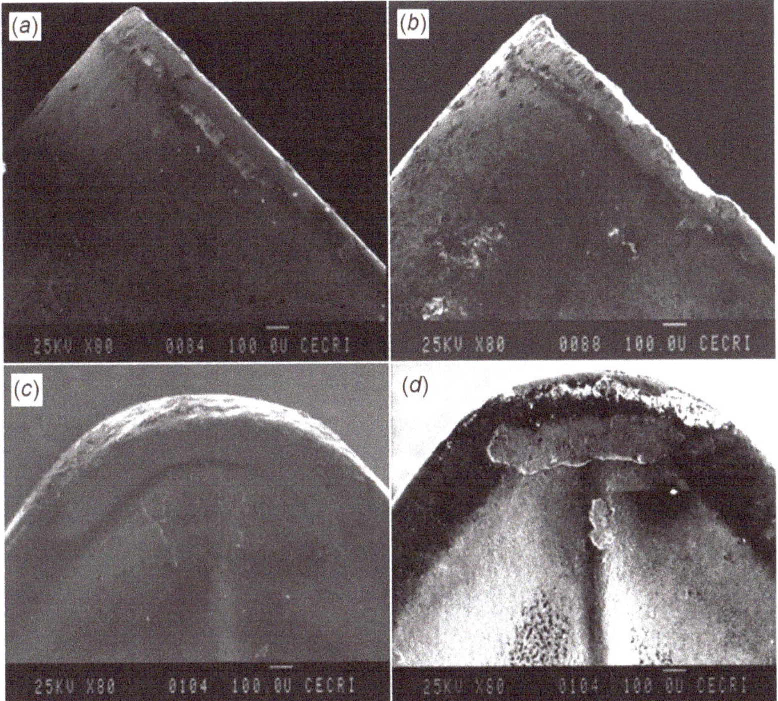

Figure 13. Flank and crater wear of cryo-treated (**a**,**b**) and non-treated (**b**,**d**) inserts [118].

SreeramaReddy et al. examined the effects of DCT on tungsten carbide cutting tool inserts performance during machining of C45 steel [119]. Cutting tool flank wear, cutting force, and surface roughness were set as indicators to evaluate the machinability of C45 and deep cryogenic treated tungsten carbide inserts. The cutting tool durability after DCT was improved as the hardness increased in comparison to untreated inserts. A better surface was obtained using a DCT treated tool as the wear resistance of the insert was increased and a negligible alteration in the cutting edge geometry was observed (Figure 14). An increment in the carbide grain size after DCT increased the thermal conductivity of the cemented carbide as a result of an increase in carbide grain contiguity and the dominant effect of the carbide phase. This impact reduces the tool tip temperature and improves its resistance against the erosion.

The machinability of coated tungsten carbide tool inserts of ISO P-40 exposed to DCT has been examined in terms of wear resistance, cutting force, and surface roughness [120]. Turning was carried out on AISI 1040 steel using untreated and cryo-treated (−176 °C) cutting tools. DCT improves the thermal conductivity of the insert and the cutting zone temperature is dissipated by the chips and consequently encourages lesser tool wear. Higher thermal conductivity of the cutting tool plays an important role at higher cutting speeds. This results in massive seizure through the chips and enables the discontinuous chips, which improves tool life by reducing the adhesive wear rate of cutting edges leading to better surface finish [121].

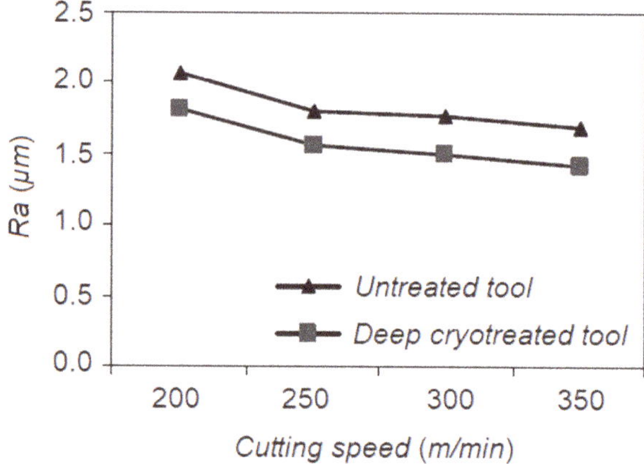

Figure 14. Surface roughness at different cutting speeds of P-30 insert [119].

Sundaram et al. applied DCT to improve the mechanical and electrical properties of copper tool electrodes, while electrical discharge machining (EDM) of Be–Cu. A higher material removal rate (MRR) was obtained by the cryo-treated electrodes due to an improvement of electrical conductivity, but the impact on the electrode wear rate (EWR) was marginal [122].

Cryogenically treated copper electrodes were employed to study the effect of DCT during the electrical discharge drilling (EDD) of Ti-6246. Hole quality in terms of surface roughness and overcut was examined to compare cryo-treated copper electrode performance with the as-received one. MRR and EWR produced surface finish, and the hole quality using cryo-treated electrodes was significantly improved [123].

Gill et al. conducted an experimental investigation to examine the effects of SCT and DCT on tungsten carbide inserts (P25) during dry turning of hot-rolled annealed steel stock (C-65). Maximum flank wear (ISO 3685-1993) and surface roughness were considered as criteria to determine the turning performance. Longer durability was recorded for the DCT insert followed by SCT tools as compared to untreated ones. Cryogenic treatment causes higher precipitation of η-phase carbides within the matrix and enhances the carbide tool life [124].

The performance of cubic boron nitride (CBN) inserts in terms of flank wear, surface finish, white layer formation, and microhardness were compared against cryo-treated coated/uncoated carbide cutting tools during dry turning of hardened AISI H11 steel. The surface roughness produced by cryo-treated inserts were comparable with that generated by CBN tools. In addition, cryogenically treated inserts provide longer tool life with a 16–23% improvement [125].

Longer tool life and 35% reduction in surface roughness (Ra) were reported using cryo-treated AISI M2 tools during drilling of mild steel C45 as results of wear resistance improvement, and a reduction in the anisotropy of the tool. Cryogenic temperature and soaking time were the most influential parameters, respectively [126].

The performance of cryogenically treated M35 HSS under dry drilling of stainless steels has been investigated. Better wear resistance and longer tool life were obtained due to the transformation of retained austenite to martensite, higher volume fraction, and more homogenous distribution of carbide particles [127].

Experimental trials were conducted to evaluate the performance of cryo-treated copper electrodes and machining EDM using a dielectric mixed with graphite powder additive while machining Inconel 718. MRR and EWR were improved using the cryo-treated copper electrode, due to the formation of hard carbide compounds on the electrode caused by the migration of carbon from the decomposed dielectric fluid and refinement of the grains, which improved the electrical and thermal conductivity [128].

The addition of metal powder in the dielectric improved the machining efficiency by making discharge breakdown easier and enlarging the discharge gap [129].

The effects of CT on surface residual stresses of ground carbide surfaces after grinding using diamond abrasive wheel were examined. Thermal residual stresses were induced due to thermal mismatch between the WC and the Co phase. A 20% reduction in residual stresses after CT execution was observed due to the cracking and plastic deformation in the WC grains [130].

A significant improvement of the cutting performance of cryo-treated M35 HSS during drilling of AISI 316 austenitic stainless steel was observed in comparison to untreated drills. Tool life, surface roughness, and hole quality were selected as criteria to assess the tool life. The improvements were attributed to the transformation of retained austenite into martensite and a more homogeneous distribution of carbides due to secondary carbide precipitation [131].

Kapoor et al. examined the performance of deep cryo-treated brass electrodes during EDM of En-31. Significant improvement of the electrical conductivity and MRR after DCT was observed due to refinement of grains and a reduction in microcavities [132]. It is worth mentioning that the electrical conductivity is proportional to the thermal conductivity, therefore, the higher the electrical conductivity, the higher the MRR [133,134].

Çiçek et al. examined the effect of CHT, CT, and CTT on the machinability of AISI H13, subjected to turning using CC670 and CC650 ceramic inserts, in terms of tool wear, surface roughness, and cutting force. A better surface quality for the CCT workpieces was obtained, followed by the CT and CHT specimens. The cryo-tempering process reduces the workpiece hardness and encourages a lower cutting force, tool wear, and surface roughness, due to the formation of higher volume fraction of fine carbides [135].

The effects of DCT on the machining efficiency of microelectric discharge machining (MEDM) of AISI 304, using copper, brass, and tungsten electrodes were evaluated. Lower EWR was obtained using cryo-treated tungsten followed by the brass and copper electrodes with a reduction of 58%, 51%, and 35%, respectively. A higher volume fraction of refined particles was observed within the brass matrix after execution of DCT as compared to copper and tungsten. DCT increases average crystal size of brass, copper, and tungsten by the value of 29%, 12%, and 4%, respectively, which results in the enhancement of hardness and wear resistance [136].

The effects of different soaking times on durability (wear resistance) of deep cryogenic treated AISI 316 austenitic stainless steel has been studied. Wear tests were conducted at different cutting speeds with a constant feed rate and depth of cut under dry turning. An improvement of the wear resistance of the cutting tool was observed. However, flank wear and crater wear were presented as a result of new η-carbide particles in the microstructure formation and a uniform and homogeneous distribution of small-sized carbide particles [137].

Mavi and Korkut examined the effect of CT on the performance of cemented carbide during turning of Ti-6Al-4V. An increment of 22% of cryo-treated inserts was reported as a consequence of fine and uniform distribution of carbide particles within the matrix. Higher thermal conductivity of the cryo-treated inserts reduces cutting zone temperature and facilitates chip removal and consequently, a lower cutting force and better surface roughness were obtained [138].

Three different grades of titanium alloy, Ti (grade two), Ti-6Al-4V (grade five), and Ti-5Al-2.5Sn (grade six) subjected to SCT and DCT, were machined by EDM. Copper, copper–chromium, and copper–tungsten electrodes were used to get a comprehensive insight into the cryogenic treatments impact. Marginal improvement of MRR was observed [139].

The effects of DCT on the machinability and wear behavior of TiAlN coated tools under dry turning of AISI 5140 have been studied [140]. These influences were examined by means of cutting force, cutting zone temperature, surface texture, and the tool life. Strengthening of the TiAlN coating on the tungsten carbide, improvement of substrate interfacial adhesion bonding, homogeneous carbide distribution, higher hardness, and better thermal conductivity were obtained after DCT. Therefore,

DCT coated inserts show better machinability in comparison to uncoated, conventionally treated cutting tools.

Wear resistance and tool life of H13A tungsten carbide inserts subjected to DCT during turning of AISI 1045 steel has been investigated. Deep cryo-treated H13A provided higher hardness, better abrasive wear resistance, and lower toughness. DCT promotes a higher volume fraction of the cobalt phase without swelling of the tungsten carbide grains and strengthens the bonding between the tungsten carbide grains and the cobalt binder, which improves inserts corrosion resistance. It was highlighted that built up edge (BUE) was the predominant mechanism of rake surface adhesion wear of the cryo-treated inserts [141].

The effects of post treatment processes, cryogenic treatment, controlled heating, and oil quenching on cutting performance of cemented tungsten carbide (K20) under turning of Inconel 718 superalloy have been examined. Higher microhardness was obtained by oil quenched treatment following CT, as the quenching increases the stresses imposed on the carbide phase (compressive stresses) and the binder (tensile stresses) with decreasing cobalt content, which reduces ductility. In contrast, CT tends to relieve the stresses induced during the sintering process that is used to produce carbide tools, which improves the hardness in comparison to untreated inserts. In addition, CT encourages more densification of the cobalt binder which is strongly bonded by tungsten carbides and improves wear resistance and reduces cutting force [142].

The influences of different treatments on grindability of 9Mn2V using a grinding wheel (3SG80KV) have been studied. Four treatment process were conducted: 1) quenching followed by oil cooling; 2) quenching followed by oil cooling and tempering; 3) quenching followed by oil cooling, CT, and tempering; and 4) quenching followed by oil cooling, CT, and two times tempering. Cryogenic treatment promotes the uniform distribution of martensite, a higher volume fraction of refined carbides, and the transformed austenite to martensite. Carbide segregation occurred in all specimens but in the last two treatments comprising CT, carbide segregation was improved after CT and tempering. Machinability of two last samples was significantly improved due to higher volume fraction and uniform distribution of secondary carbides. In addition, impact toughness improvement and residual stresses relief on the workpieces surface due to CT and tempering, especially for the cryo-treated sample followed by double tempering, were detected [143].

The cutting capabilities of M2 HSS drill subjected to CT, CTT, and TiAlN/TiN-coated were evaluated while drilling Ti-6Al-4V under dry and wet conditions. Cryo-treated drills show higher wear resistance as compared to untreated and TiAlN/TiN-coated. After CT execution, the tool microstructure received finer carbide particles with a homogenous distribution, as well as a higher volume fraction of martensite, which encourages longer tool life in comparison to untreated drills. Cryogenic treatment is more cost effective than coating and brings fairly remarkable improvements [144].

Özbek et al. evaluated uncoated and deep cryo-treated tungsten carbide insert's performance while turning an austenitic stainless steel (AISI 316). Flank, notch, and crater wear were selected as indicators to examine the tool life. A higher volume fraction of fine η-carbides after DCT execution, improved the hardness and wear resistance. DCT encourages larger grain size and better thermal conductivity which improved the flank wear by 34% and the crater wear by 53% [145].

Xu et al. examined the effects of DCT on the residual stresses and mechanical properties of electron beam (EB)-welded Ti-6Al-4V (TC4) joints [146]. Soaking time influence was evaluated to determine an optimum time. It was highlighted that a reduction in longitudinal and transversal residual stress (with respect to welding direction) after DCT was different. It was observed that the hardness in the welded area was higher than that in the base metal, and the average values of hardness increased with DCT due to the alteration of quantity, size, and morphology of the α and β phases.

Khanna and Singh studied the impact of DCT on machining efficiency using a high chromium cold alloy tool (D3) during WEDM (wire-cut electrical discharge machining). MRR was reduced by 5.6% after DCT due to an improvement of thermal conductivity as result of the transformation of retained austenite to martensite and the refinement of carbide particles, which promotes the heat flow laterally across the line of the cut [147].

The machinability of the AISI D2 tool steel subjected to DCT during EDM using copper electrodes has been investigated. Higher MRR (~18% enhancement), lower EWR, and surface roughness (26% and 11% improvement, respectively) were obtained after DCT as compared to untreated workpieces [148].

The performance of cutting inserts PVD AlTiN-coated (KC5525) and uncoated (K313) subjected to the DCT during dry turning of Nimonic 90 alloy were investigated. DCT promotes the formation of η-phase carbide particles and grain refinement which results in higher hardness and wear resistance. In addition, DCT strengthens the coating and reduces the failure probability in comparison to coating durability and damage of untreated inserts [149].

Khan et al. used cryo-treated K313 inserts during dry turning of CP-Ti grade two to evaluate the effects of DCT on tool performance. DCT increased the microhardness, wear resistance, and improved the chip formation phenomena such as chip compression ratio and friction coefficient. An increment of η-phase carbides is to be recognized as the main contributor to these improvements [150].

Naveena et al. carried out experimental trials to evaluate cryogenically treated tungsten carbide under drilling of AISI 304. Surface roughness using cryo-treated tools was reduced by 40% due to the enhancement of wear resistance and cutting edge stability. DCT encouraged a 19% reduction in average grain size of the α-phase and consequently, increased the hardness. Furthermore, DCT reduced the volume fraction of β-phase Co which implied combination of cobalt with WC particles to form η-phase carbides [151].

The research works devoted to examining the effects of CT on machinability of different materials subjected to nonconventional and conventional manufacturing process have been tabulated by Tables 4 and 5, respectively.

Table 4. Literature data related applying cryogenic treatment in nonconventional manufacturing process.

First Author, [#]	Process	Tool Material	Workpiece Material	Key Findings
Sundaram [122]	EDM	Copper	Be–Cu	Higher MRR due to electrical conductivity improvement, and marginal effect on electrode life
Gill [123]	EDD	Copper	Ti-6246	DCT improves MRR, electrode life, and surface finish.
Kumar [128]	EDM	Copper	Inconel 718	MRR and EWR have been improved due to formation of hard carbide compounds on the electrode.
Kapoor [132]	EDM	Brass	En-31	Improvement of electrical conductivity and MRR after DCT has been achieved due to refinement of grains and reduction in microcavities.
Jafferson [136]	MEDM	Copper	AISI 304	DCT encourages average crystal size of brass, copper, and tungsten by the value of 29%, 12%, and 4%, respectively, which result in enhancement of hardness and wear resistance.
Kumar [139]	EDM	Copper and Copper-Tungsten	Ti, Ti-6Al-4V, and Ti-5Al-2.5Sn	Marginal improvement of MRR was observed.
Xu [146]	EB welding	N.A.	Ti-6Al-4V	Due to alteration of quantity, size, and morphology of the α and β phases after DCT, hardness in welded area was higher than that in the base metal.
Khanna [147]	WEDM	N.A.	AISI D3	Transformation of retained austenite to martensite and refinement of carbide particles after DCT execution, thermal conductivity has been improved.
Goyal [148]	EDM	Copper	AISI D2	Higher MRR (~18% enhancement), lower EWR, and surface roughness (26% and 11% improvement, respectively) were obtained after DCT.

Table 5. Literature data related applying cryogenic treatment in conventional manufacturing process.

First Author, [#]	Process	Tool Material	Workpiece Material	Key Findings
Yong [116]	Milling	Tungsten carbide G10E	ASSAB 760	DCT improves tool life as consequence of higher heat transfer.
Firouzdor [117]	Drilling	M2 HSS	CK40	Better wear resistance against diffusion wear.
Vadivel [118]	Turning	Coated carbide inserts	AISI/SAE 80-55-06 SG	Higher hardness and better wear performance due to uniform distribution and higher volume fraction of refine carbides particles.
SreeramaReddy [119]	Turning	Tungsten carbide	C45 steel	Increment of carbide grain size after DCT increased the thermal conductivity and reduced cutting tool tip temperature.
Reddy [120]	Turning	ISO P-40	AISI 1040	Lower tool wear due to thermal conductivity improvement after DCT has been observed.
Gill [124]	Turning	P25	C-65	Longer durability was recorded for DCT insert followed by SCT due to higher precipitation of η-phase carbides.
Dogra [125]	Turning	Cubic boron nitride (CBN)	AISI H11	16%–23% tool life improvement has been reported.
Shirbhate [126]	Drilling	AISI M2	C45	Longer tool life and 35% reduction in surface roughness (Ra) were reported after DCT execution.
Çiçek [127]	Drilling	M35 HSS	Stainless steels	Longer tool life was obtained due to transformation of retained austenite to martensite and homogenous distribution of carbides particles.
Yuan [130]	Grinding	Diamond abrasive wheel	Ultra-fine grade cemented carbide	20% reduction in residual stresses after CT execution were observed due to the cracking and plastic deformation in the WC grains.
Çiçek [131]	Drilling	M35 HSS	AISI 316	Transformation of retained austenite into martensite and more homogeneous distribution of carbides provided better tool performance.
Çiçek [133]	Turning	Ceramic Inserts	AISI H13	DCT reduces tool wear and surface roughness as result of higher volume fraction of fine carbides formation.
Özbek [137]	Turning	Cemented carbide	AISI 316	Tool life was improved due to homogeneous distribution of small-sized carbide particles.
Mavi [138]	Turning	Cemented carbide	Ti-6Al-4 V	Tool life has been improved by 22% due to higher thermal conductivity improvement.
He [140]	Turning	Tungsten carbide	AISI 5140	DCT coated inserts were shown better machinability in terms of cutting force, cutting zone temperature, surface texture, and tool life.
Thornton [141]	Turning	H13A	AISI 1045	Better corrosion resistance obtained due to strengthen of carbide grains and the cobalt binder.
Thakur [142]	Turning	K 20	Inconel 718	CT encourages more densification of the cobalt binder which is strongly bonded by tungsten carbides and improves tool wear resistance.
Dong [143]	Grinding	Grinding wheel (3SG80KV)	9Mn2 V	Improvement and releasement of residual stresses on the workpieces surface has been improved.
Kivak [144]	Drilling	M2 HSS	Ti-6Al-4 V	It was concluded that CT is more cost effective than coating which brings remarkable improvements.
Özbek [145]	Turning	Tungsten carbide	AISI 316	Higher volume fraction of fine η-carbides after DCT execution improves the hardness and wear resistance.
Chetan [149]	Turning	KC5525 and K313	Nimonic 90	DCT strengthens the coating and reduces failure probability in comparison to coating durability and damage on untreated inserts
Khan [150]	Turning	K313	CP-Ti grade 2	DCT increases microhardness, wear resistance and improves chip formation phenomenon.
Naveena [151]	Drilling	Tungsten Carbide	AISI 304	DCT encourages 19% reduction in average grain size of α-phase and consequently increases the hardness and improves wear resistance.

7. Conclusions

A literature survey has been conducted to focus on cryogenic treatments effects on microstructural alteration and its correlation with material mechanical properties and manufacturability. Comparative investigations have been made to evaluate the effects of CHT, SCT, and DCT on the microstructure and mechanical property variations. The effects of important parameters and their interactions such as austenitizing temperatures, ramp down, soaking time, and temperature, as well as ramp up, are discussed. The review of the literature implies that the evaluation of microstructural variations due to cryo-treatment and their interaction with mechanical properties is a complex phenomenon. Therefore, these complexities of interactions of microstructural alteration and mechanical properties depend on the adopted methods such as prior and supplementary treatments (austenitizing, quenching, and tempering), influential parameters, and demanded accuracy.

A review of the literature and the synthesis of technical contributions indicates various ways to execute cryogenic treatments. The important conclusions of this literature survey and the future directions are suggested as follows:

- In almost all steels, elimination of retained austenite, the increment of carbide volume fraction with homogenous distribution, and uniform size, as well as the formation of fine secondary carbide precipitation, are the greatest contributor to hardness improvement after DCT. Microstructure during SCT experiences these changes with a lower magnitude in comparison to DCT, but more efficiently than CHT;

- Uniform carbide distribution and homogeneous particle size, along with the formation of new small size carbides, have been obtained after DCT which result in higher wear resistance. Meanwhile, the formation of η-carbide and a homogeneous distribution of the produced carbides are the most important players in wear resistance improvement, rather than only the elimination of retained austenite. In the case of CT of steels, to obtain better properties and to improve treatment efficiency, it is recommended that austenizing, quenching, DCT, and tempering are carried out one after another in a cycle;

- Homogenous distribution of secondary carbides and the transformation of retained austenite to martensite after SCT and DCT improve fatigue strength and tensile toughness;

- Applying appropriate cooling temperature and soaking time causes substantial transformation of the soft austenite into hard martensite and in addition, metallurgical alteration of the martensite can be obtained;

- It is relevant to note that martensitic transformation never goes to completion after DCT, and the retained austenite always exists in the structure of high-carbon martensite. However, further tempering is suggested to boost the elimination of austenite and the formation of secondary carbides within the martensite;

- DCT encourages the increment of carbide grain size and improves the thermal conductivity of cemented carbide cutting tools. Therefore, as consequence of the increment in carbide grain contiguity, and the dominant effect of the carbide phase, higher wear resistance, and better surface roughness can be obtained;

- It is recommended to standardize the CT cycles including ramp down, soaking time and temperature as well as ramp up and supplementary processes, to optimize the properties of each specific material. Process qualification and standards would be drafted and validated for more promising alloys;

- The effects of CT on different cutting tools needs to be clarified and standardized to improve the performance of inserts in terms of wear, hardness, dimensional stability, and thermal conductivity.

- Special efforts are required to characterize the correlation between microstructural alterations and wear mechanisms of cutting tools from different materials.

Funding: This research received no external funding.

Acknowledgments: The authors wish to gratefully thank DEPEC group.

Conflicts of Interest: The authors declare no conflict of interest.

Nomenclature

Conventional heat treatment	CHT
Cryogenic treatment	CT
Cryogenic treatment followed by tempering	CTT
Cryogenic-ageing circular treatment	CACT
Deep cryogenic treatment	DCT
Deep cryogenic treatment pluse tempering	DCTT
Differential scanning calorimetry	DSC
Electron back scattered diffraction	EBSD
Electrical discharge machining	EDM
Electrode wear rate	EWR
Electric discharge drilling	EDD
Friction Stir Welding	FSW
Microelectric discharge machining	MEDM
Material removal rate	MRR
Maximum Likelihood Estimation	MLE
Optical microscopy	OM
Subzero treatment	SZT
Shallow cryogenic treatment	SCT
Scanning electron microscopy	SEM
Transmission electron microscopy	TEM
Tungsten Inert Gas Welding	TIG
Wire-cut electrical discharge machining	WEDM
X-ray diffraction	XRD

References

1. Pellizzari, M. Influence of deep cryogenic treatment on heat treatment of steel and Cu–Be alloy. *Int. Heat Treat. Surf. Eng.* **2010**, *4*, 105–109. [CrossRef]
2. Kalia, S. Cryogenic Processing: A Study of Materials at Low Temperatures. *J. Low Temp. Phys.* **2010**, *158*, 934–945. [CrossRef]
3. Das, D.; Dutta, A.K.; Toppo, V.; Ray, K.K. Effect of Deep Cryogenic Treatment on the Carbide Precipitation and Tribological Behavior of D2 Steel. *Mater. Manuf. Process.* **2007**, *22*, 474–480. [CrossRef]
4. Surberg, C.H.; Stratton, P.; Lingenhöle, K. The effect of some heat treatment parameters on the dimensional stability of AISI D2. *Cryogenics* **2008**, *48*, 42–47. [CrossRef]
5. Bensely, A.; Prabhakaran, A.; Lal, D.M.; Nagarajan, G. Enhancing the wear resistance of case carburized steel (En 353) by cryogenic treatment. *Cryogenics* **2005**, *45*, 747–754. [CrossRef]
6. Chopra, S.A.; Sargade, V.G. Metallurgy behind the Cryogenic Treatment of Cutting Tools: An Overview. *Mater. Today* **2015**, *2*, 1814–1824. [CrossRef]
7. Das, D.; Dutta, A.K.; Ray, K.K. Sub-zero treatments of AISI D2 steel: Part II. Wear behavior. *Mater. Sci. Eng. A* **2010**, *527*, 2194–2206. [CrossRef]
8. Bensely, A.; Venkatesh, S.; Lal, D.M.; Nagarajan, G.; Rajadurai, A.; Junik, K. Effect of cryogenic treatment on distribution of residual stress in case carburized En 353 steel. *Mater. Sci. Eng. A* **2008**, *479*, 229–235. [CrossRef]
9. Ghosh, S.; Rao, P.V. Application of sustainable techniques in metal cutting for enhanced machinability: A review. *J. Clean. Prod.* **2015**, *100*, 17–34. [CrossRef]
10. Jayal, A.D.; Badurdeen, F.; Dillon, O.W., Jr.; Jawahir, I.S. Sustainable manufacturing: Modeling and optimization challenges at the product, process and system levels. *CIRP J. Manuf. Sci. Technol.* **2010**, *2*, 144–152. [CrossRef]

11. Gill, S.S.; Singh, J.; Singh, H.; Singh, R. Investigation on wear behaviour of cryogenically treated TiAlN coated tungsten carbide inserts in turning. *Int. J. Mach. Tools Manuf.* **2011**, *51*, 25–33. [CrossRef]

12. Das, D.; Ray, K.K.; Dutta, A.K. Influence of temperature of sub-zero treatments on the wear behaviour of die steel. *Wear* **2009**, *267*, 1361–1370. [CrossRef]

13. Diekman, F. Cold and Cryogenic Treatment of Steel. In *ASM Handbook*; ASM International: Novelty, OH, USA, 2013; pp. 382–386.

14. Baldissera, P.; Delprete, C. Deep cryogenic treatment: A bibliographic review. *Open Mech. Eng. J.* **2008**, *2*, 1–11. [CrossRef]

15. Thornton, R.; Slatter, T.; Ghadbeigi, H. Effects of deep cryogenic treatment on the dry sliding wear performance of ferrous alloys. *Wear* **2013**, *305*, 177–191. [CrossRef]

16. Podgornik, B.; Leskovšek, V.; Vižintin, J. Deep-Cryogenic Treatment and Effect of Austenizing Temperature on Tribological Performance of P/M High-Speed Steel. In Proceedings of the 9th Biennial Conference on Engineering Systems Design and Analysis, ASME, Haifa, Israel, 7–9 July 2008; pp. 509–512. [CrossRef]

17. Cajner, F.; Landek, D.; Rafael, H.; Šolić, S.; Kovačić, S. Effect of deep cryogenic treatment on dilatometric curve and tribological properties of high speed steel. *Int. Heat Treat. Surf. Eng.* **2012**, *6*, 67–71. [CrossRef]

18. Gill, S.S.; Singh, H.; Singh, R.; Singh, J. Cryoprocessing of cutting tool materials—A review. *Int. J. Adv. Manuf. Technol.* **2010**, *48*, 175–192. [CrossRef]

19. Kalpakjian, S.; Schmid, S.R. *Manufacturing Engineering and Technology*; Pearson Education South Asia Pte Ltd.: Singapore, 2014.

20. Kumar, T.V.; Thirumurugan, R.; Viswanath, B. Influence of cryogenic treatment on the metallurgy of ferrous alloys: A review. *Mater. Manuf. Process.* **2017**, *32*, 1789–1805. [CrossRef]

21. Papp, R. The Emerging Science of Deep Cryogenic Treatment. *Cold Facts* **2014**, *30*, 8–10.

22. Wood, R. Classic papers: Cryogenic heat treatment. *Int. Heat Treat. Surf. Eng.* **2008**, *2*, 74–75. [CrossRef]

23. Amini, K.; Akhbarizadeh, A.; Javadpour, S. Cryogenic heat treatment of the ferrous materials—A review of the current state. *Met. Res. Technol.* **2016**, *113*, 611. [CrossRef]

24. Wu, Z. The Expanded Application Research of Deep Cryogenic Treatment. In Proceedings of the IEEE 2009 International Conference on Measuring Technology and Mechatronics Automation, Zhangjiajie, China, 11–12 April 2009. [CrossRef]

25. Gogte, C.L.; Iyer, K.M.; Paretkar, R.K.; Peshwe, D.R. Deep subzero processing of metals and alloys: Evolution of microstructure of AISI T42 tool steel. *Mater. Manuf. Process.* **2009**, *24*, 718–722. [CrossRef]

26. Diekman, R. Deep Cryogenic Treatment. *Therm. Process. Gear Solut.* **2013**, 52–55. Available online: http://thermalprocessing.com/deep-cryogenic-treatment/ (accessed on 12 September 2019).

27. Das, D.; Dutta, A.K.; Ray, K.K. Correlation of microstructure with wear behaviour of deep cryogenically treated AISI D2 steel. *Wear* **2009**, *267*, 1371–1380. [CrossRef]

28. Ray, K.K.; Das, D. Improved wear resistance of steels by cryotreatment: The current state of understanding. *Mater. Sci. Tech.* **2017**, *33*, 340–354. [CrossRef]

29. Wang, J.; Fu, R.; Li, Y.; Zhang, J. Effects of deep cryogenic treatment and low-temperature aging on the mechanical properties of friction-stir-welded joints of 2024-T351 aluminum alloy. *Mater. Sci. Eng. A* **2014**, *609*, 147–153. [CrossRef]

30. Gill, S.S.; Singh, J.; Singh, R.; Singh, H. Metallurgical principles of cryogenically treated tool steels—A review on the current state of science. *Int. J. Adv. Manuf. Technol.* **2011**, *54*, 59–82. [CrossRef]

31. Tated, R.G.; Patil, P.I. Comparison of effects of cryogenic treatment on different types of steels: A review. In *Proceedings of the IJCA International Conference in Computational Intelligence 9, 10–29 March 2012*; IJCA Journal: St. Gallen, Switzerland, 2012.

32. Huang, J.Y.; Zhu, Y.T.; Liao, X.Z.; Beyerlein, I.J.; Bourke, M.A.; Mitchell, T.E. Microstructure of cryogenic treated M2 tool steel. *Mater. Sci. Eng. A* **2003**, *339*, 241–244. [CrossRef]

33. Sachin, S.S. Cryogenic hardening and its effects on properties of an aerospace aluminium alloy. *Int. J. Late. Trend. Eng. Tech.* **2017**, *8*, 566–571. [CrossRef]

34. Amini, K.; Akhbarizadeh, A.; Javadpour, S. Investigating the effect of holding duration on the microstructure of 1.2080 tool steel during the deep cryogenic heat treatment. *Vacuum* **2012**, *86*, 1534–1540. [CrossRef]

35. Li, S.; Deng, L.; Wu, X.; Wang, H.; Min, Y.A.; Min, N. Effect of deep cryogenic treatment on internal friction behaviors of cold work die steel and their experimental explanation by coupling model. *Mater. Sci. Eng. A* **2010**, *527*, 7950–7954. [CrossRef]

36. Li, S.; Min, N.; Li, J.; Wu, X.; Li, C.; Tang, L. Experimental verification of segregation of carbon and precipitation of carbides due to deep cryogenic treatment for tool steel by internal friction method. *Mater. Sci. Eng. A* **2013**, *575*, 51–60. [CrossRef]

37. Leskovsek, V.; Ule, B. Influence of deep cryogenic treatment on microstructure, mechanical properties and dimensional changes of vacuum heat-treated high-speed steel. *Heat Treat. Met.* **2002**, *29*, 72–76. [CrossRef]

38. Gao, Y.; Luo, B.H.; Bai, Z.H.; Zhu, B.; Ouyang, S. Effects of deep cryogenic treatment on the microstructure and properties of WCFeNi cemented carbides. *Int. J. Refract. Metal. Hard Mater.* **2016**, *58*, 42–50. [CrossRef]

39. Zhirafar, S.; Rezaeian, A.; Pugh, M. Effect of cryogenic treatment on the mechanical properties of 4340 steel. *J. Mater. Process. Technol.* **2007**, *186*, 298–303. [CrossRef]

40. Casagrande, A.; Cammarota, G.P.; Micele, L. Relationship between fatigue limit and Vickers hardness in steels. *Mater. Sci. Eng. A* **2011**, *528*, 3468–3473. [CrossRef]

41. Vimal, A.J.; Bensely, A.; Lal, D.M.; Srinivasan, K. Deep cryogenic treatment improves wear resistance of En 31 steel. *Mater. Manuf. Process.* **2008**, *23*, 369–376. [CrossRef]

42. Harish, S.; Bensely, A.; Lal, D.M.; Rajadurai, A. Microstructural study of cryogenically treated En 31 bearing steel. *J. Mater. Process. Technol.* **2009**, *209*, 3351–3357. [CrossRef]

43. Li, S.; Xie, Y.; Wu, X. Hardness and toughness investigations of deep cryogenic treated cold work die steel. *Cryogenics* **2010**, *50*, 89–92. [CrossRef]

44. Jeleńkowski, J.; Ciski, A.; Babul, T. Effect of deep cryogenic treatment on substructure of HS6-5-2 high speed steel. *J. Achie. Mater. Manuf. Eng.* **2010**, *43*, 80–87.

45. Senthilkumar, D.; Rajendran, I.; Pellizzari, M.; Siiriainen, J. Influence of shallow and deep cryogenic treatment on the residual state of stress of 4140 steel. *J. Mater. Process. Technol.* **2011**, *211*, 396–401. [CrossRef]

46. El Mehtedi, M.; Ricci, P.; Drudi, L.; El Mohtadi, S.; Cabibbo, M.; Spigarelli, S. Analysis of the effect of deep cryogenic treatment on the hardness and microstructure of X30 CrMoN 15 1 steel. *Mater. Des.* **2012**, *33*, 136–144. [CrossRef]

47. Li, J.; Tang, L.; Li, S.; Wu, X. Finite element simulation of deep cryogenic treatment incorporating transformation kinetics. *Mater. Des.* **2013**, *47*, 653–666. [CrossRef]

48. Candane, D.; Alagumurthi, N.; Palaniradja, K. Effect of cryogenic treatment on microstructure and wear characteristics of AISI M35 HSS. *Int. J. Mater. Sci. Appl.* **2013**, *2*, 56–65. [CrossRef]

49. Akhbarizadeh, A.; Javadpour, S. Investigating the effect of as-quenched vacancies in the final microstructure of 1.2080 tool steel during the deep cryogenic heat treatment. *Mater. Lett.* **2013**, *93*, 247–250. [CrossRef]

50. SreeramaReddy, T.V.; Sornakumar, T.; VenkataramaReddy, M.; Venkatram, R. Machining performance of low temperature treated P-30 tungsten carbide cutting tool inserts. *Cryogenics* **2008**, *48*, 458–461. [CrossRef]

51. Zhang, H.; Chen, L.; Sun, J.; Wang, W.; Wang, Q. Influence of Deep Cryogenic Treatment on Microstructures and Mechanical Properties of an Ultrafine-Grained WC–12Co Cemented Carbide. *Acta Met. Sin.* **2014**, *27*, 894–900. [CrossRef]

52. Idayan, A.; Gnanavelbabu, A.; Rajkumar, K. Influence of deep cryogenic treatment on the mechanical properties of AISI 440C bearing steel. *Procedia Eng.* **2014**, *97*, 1683–1691. [CrossRef]

53. Xie, C.H.; Huang, J.W.; Tang, Y.F.; Gu, L.N. Effects of deep cryogenic treatment on microstructure and properties of WC-11Co cemented carbides with various carbon contents. *Trans. Nonferr. Metal. Soc.* **2015**, *25*, 3023–3028. [CrossRef]

54. Yuan, C.; Wang, Y.; Sang, D.; Li, Y.; Jing, L.; Fu, R.; Zhang, X. Effects of deep cryogenic treatment on the microstructure and mechanical properties of commercial pure zirconium. *J. Alloy Compd.* **2015**, *619*, 513–519. [CrossRef]

55. Pérez, M.; Belzunce, F.J. The effect of deep cryogenic treatments on the mechanical properties of an AISI H13 steel. *Mater. Sci. Eng. A* **2015**, *624*, 32–40. [CrossRef]

56. Mohan, K.; Suresh, J.A.; Ramu, P.; Jayaganthan, R. Microstructure and Mechanical Behavior of Al 7075-T6 Subjected to Shallow Cryogenic Treatment. *J. Mater. Eng. Perform.* **2016**, *25*, 2185–2194. [CrossRef]

57. Nazarian, H.; Krol, M.; Pawlyta, M.; Vahdat, S.E. Effect of sub-zero treatment on fatigue strength of aluminum 2024. *Mater. Sci. Eng. A* **2018**, *710*, 38–46. [CrossRef]

58. Taşkesen, A.; Aksöz, S.; Özdemir, A.T. Effect of Sub-Zero Treatment on Ageing of Aluminium Composite. *Int. J. Mater. Mech. Manuf.* **2018**, *6*, 26–30. [CrossRef]

59. Thakur, D.; Ramamoorthy, B.; Vijayaraghavan, L. Influence of different post treatments on tungsten carbide–cobalt inserts. *Mater. Lett.* **2008**, *62*, 4403–4406. [CrossRef]

60. Akhbarizadeh, A.; Golozar, M.A.; Shafeie, A.; Kholghy, M. Effects of austenizing time on wear behavior of D6 tool steel after deep cryogenic treatment. *J. Iron Steel Res. Int.* **2009**, *16*, 29–32. [CrossRef]

61. Dhokey, N.B.; Nirbhavne, S. Dry sliding wear of cryotreated multiple tempered D-3 tool steel. *J. Mater. Process. Technol.* **2009**, *209*, 1484–1490. [CrossRef]

62. Akhbarizadeh, A.; Shafyei, A.; Golozar, M.A. Effects of cryogenic treatment on wear behavior of D6 tool steel. *Mater. Des.* **2009**, *30*, 3259–3264. [CrossRef]

63. Das, D.; Dutta, A.K.; Ray, K.K. Optimization of the duration of cryogenic processing to maximize wear resistance of AISI D2 steel. *Cryogenics* **2009**, *49*, 176–184. [CrossRef]

64. Straffelini, G.; Bizzotto, G.; Zanon, V. Improving the wear resistance of tools for stamping. *Wear* **2010**, *269*, 693–697. [CrossRef]

65. Podgornik, B.; Leskovšek, V.; Vižintin, J. Influence of Deep-Cryogenic Treatment on Tribological Properties of P/M High-Speed Steel. *Mater. Manuf. Process.* **2009**, *24*, 734–738. [CrossRef]

66. Wang, J.; Xiong, J.; Fan, H.; Yang, H.S.; Liu, H.H.; Shen, B.L. Effects of high temperature and cryogenic treatment on the microstructure and abrasion resistance of a high chromium cast iron. *J. Mater. Process. Technol.* **2009**, *209*, 3236–3240. [CrossRef]

67. Oppenkowski, A.; Weber, S.; Theisen, W. Evaluation of factors influencing deep cryogenic treatment that affect the properties of tool steels. *J. Mater. Process. Technol.* **2010**, *210*, 1949–1955. [CrossRef]

68. Das, D.; Sarkar, R.; Dutta, A.K.; Ray, K.K. Influence of sub-zero treatments on fracture toughness of AISI D2 steel. *Mater. Sci. Eng.* **2010**, *528*, 589–603. [CrossRef]

69. Cajner, F.; Leskovšek, V.; Landek, D.; CaJner, H. Effect of Deep-Cryogenic Treatment on High Speed Steel Properties. *Mater. Manuf. Process.* **2009**, *24*, 743–746. [CrossRef]

70. Asl, K.M.; Tari, A.; Khomamizadeh, F. Effect of deep cryogenic treatment on microstructure, creep and wear behaviors of AZ91 magnesium alloy. *Mater. Sci. Eng.* **2009**, *523*, 27–31. [CrossRef]

71. Sogalad, I.; Udupa, N.S. Influence of cryogenic treatment on load bearing ability of interference fitted assemblies. *Mater. Des.* **2010**, *31*, 564–569. [CrossRef]

72. Akhbarizadeh, A.; Amini, K.; Javadpour, S. Effects of applying an external magnetic field during the deep cryogenic heat treatment on the corrosion resistance and wear behavior of 1.2080 tool steel. *Mater. Des.* **2012**, *41*, 114–123. [CrossRef]

73. Siva, R.S.; Jaswin, M.A.; Lal, D.M. Enhancing the Wear Resistance of 100Cr6 Bearing Steel Using Cryogenic Treatment. *Tribol. Trans.* **2012**, *55*, 387–393. [CrossRef]

74. Xu, N.; Cavallaro, G.P.; Gerson, A.R. Synchrotron micro-diffraction analysis of the microstructure of cryogenically treated high performance tool steels prior to and after tempering. *Mater. Sci. Eng. A* **2010**, *527*, 6822–6830. [CrossRef]

75. Arockia Jaswin, M.; Mohan Lal, D. Optimization of the Cryogenic Treatment Process for En 52 Valve Steel Using the Grey-Taguchi Method. *Mater. Manuf. Process.* **2010**, *25*, 842–850. [CrossRef]

76. Li, S.; Deng, L.; Wu, X.; Min, Y.A.; Wang, H. Influence of deep cryogenic treatment on microstructure and evaluation by internal friction of a tool steel. *Cryogenics* **2010**, *50*, 754–758. [CrossRef]

77. Koneshlou, M.; Asl, K.M.; Khomamizadeh, F. Effect of cryogenic treatment on microstructure, mechanical and wear behaviors of AISI H13 hot work tool steel. *Cryogenics* **2011**, *51*, 55–61. [CrossRef]

78. Senthilkumar, D.; Rajendran, I. Influence of Shallow and Deep Cryogenic Treatment on Tribological Behavior of En 19 Steel. *J. Iron. Steel. Res. Int.* **2011**, *18*, 53–59. [CrossRef]

79. Jaswin, M.A.; Lal, D.M.; Rajadurai, A. Effect of Cryogenic Treatment on the Microstructure and Wear Resistance of X45Cr9Si3 and X53Cr22Mn9Ni4N Valve Steels. *Tribol. Trans.* **2011**, *54*, 341–350. [CrossRef]

80. Li, S.; Min, N.; Deng, L.; Wu, X.; Min, Y.A.; Wang, H. Influence of deep cryogenic treatment on internal friction behavior in the process of tempering. *Mater. Sci. Eng. A* **2011**, *528*, 1247–1250. [CrossRef]

81. Amini, K.; Nategh, S.; Shafyei, A.; Rezaeian, A. Effect of deep cryogenic treatment on the properties of 80CrMo12 5 tool steel. *Int. J. Miner. Met. Mater.* **2012**, *19*, 30–37. [CrossRef]

82. Gill, S.S.; Singh, J.; Singh, R.; Singh, H. Effect of Cryogenic Treatment on AISI M2 High Speed Steel: Metallurgical and Mechanical Characterization. *J. Mater. Eng. Perform.* **2012**, *21*, 1320–1326. [CrossRef]

83. Sri Siva, R.; Mohan Lal, D.; Jaswin, M.A. Optimization of Deep Cryogenic Treatment Process for 100Cr6 Bearing Steel Using the Grey-Taguchi Method. *Tribol. Trans.* **2012**, *55*, 854–862. [CrossRef]

84. Gavriljuk, V.G.; Theisen, W.; Sirosh, V.V.; Polshin, E.V.; Kortmann, A.; Mogilny, G.S.; Petrov, Y.N.; Tarusin, Y.V. Low-temperature martensitic transformation in tool steels in relation to their deep cryogenic treatment. *Acta Mater.* **2013**, *61*, 1705–1715. [CrossRef]

85. Das, D.; Ray, K.K. Structure–property correlation of sub-zero treated AISI D2 steel. *Mater. Sci. Eng. A* **2012**, *541*, 45–60. [CrossRef]

86. Amini, K.; Akhbarizadeh, A.; Javadpour, S. Effect of deep cryogenic treatment on the formation of nano-sized carbides and the wear behavior of D2 tool steel. *Int. J. Min. Met. Mater.* **2012**, *19*, 795–799. [CrossRef]

87. Das, D.; Ray, K.K. On the mechanism of wear resistance enhancement of tool steels by deep cryogenic treatment. *Philos. Mag. Lett.* **2012**, *92*, 295–303. [CrossRef]

88. Podgornik, B.; Majdic, F.; Leskovsek, V.; Vizintin, J. Improving tribological properties of tool steels through combination of deep-cryogenic treatment and plasma nitriding. *Wear* **2012**, *288*, 88–93. [CrossRef]

89. Vahdat, S.E.; Nategh, S.; Mirdamadi, S. Microstructure and tensile properties of 45WCrV7 tool steel after deep cryogenic treatment. *Mater. Sci. Eng. A* **2013**, *585*, 444–454. [CrossRef]

90. Senthilkumar, D.; Rajendran, I. Optimization of Deep Cryogenic Treatment to Reduce Wear Loss of 4140 Steel. *Mater. Manuf. Process.* **2012**, *27*, 567–572. [CrossRef]

91. Amini, K.; Akhbarizadeh, A.; Javadpour, S. Investigating the effect of quench environment and deep cryogenic treatment on the wear behavior of AZ91. *Mater. Des.* **2014**, *54*, 154–160. [CrossRef]

92. Gunes, I.; Cicek, A.; Aslantas, K.; Kara, F. Effect of Deep Cryogenic Treatment on Wear Resistance of AISI 52100 Bearing Steel. *Trans. Indian Inst. Met.* **2014**, *67*, 909–917. [CrossRef]

93. Khun, N.W.; Liu, E.; Tan, A.W.Y.; Senthilkumar, D.; Albert, B.; Lal, D.M. Effects of deep cryogenic treatment on mechanical and tribological properties of AISI D3 tool steel. *Friction* **2015**, *3*, 234–242. [CrossRef]

94. Çiçek, A.; Kara, F.; Kıvak, T.; Ekici, E.; Uygur, I. Effects of deep cryogenic treatment on the wear resistance and mechanical properties of AISI H13 hot-work tool steel. *J. Mater. Eng. Perform.* **2015**, *24*, 4431–4439. [CrossRef]

95. Li, H.; Tong, W.; Cui, J.; Zhang, H.; Chen, L.; Zuo, L. The influence of deep cryogenic treatment on the properties of high-vanadium alloy steel. *Mater. Sci. Eng. A* **2016**, *662*, 356–362. [CrossRef]

96. Podgornik, B.; Paulin, I.; Zajec, B.; Jacobson, S.; Leskovšek, V. Deep cryogenic treatment of tool steels. *J. Mater. Process. Technol.* **2016**, *229*, 398–406. [CrossRef]

97. Li, G.R.; Cheng, J.F.; Wang, H.M.; Li, C.Q. The influence of cryogenic-aging circular treatment on the microstructure and properties of aluminum matrix composites. *J. Alloy. Compd.* **2017**, *695*, 1930–1945. [CrossRef]

98. Bensely, A.; Shyamala, L.; Harish, S.; Lal, D.M.; Nagarajan, G.; Junik, K.; Rajadurai, A. Fatigue behaviour and fracture mechanism of cryogenically treated En 353 steel. *Mater. Des.* **2009**, *30*, 2955–2962. [CrossRef]

99. Baldissera, P. Deep cryogenic treatment of AISI 302 stainless steel: Part I—Hardness and tensile properties. *Mater. Des.* **2010**, *31*, 4725–4730. [CrossRef]

100. Baldissera, P.; Delprete, C. Deep cryogenic treatment of AISI 302 stainless steel: Part II—Fatigue and corrosion. *Mater. Des.* **2010**, *31*, 4731–4737. [CrossRef]

101. Baldissera, P.; Cavalleri, S.; Marcassoli, P.; Tordini, F. Study of the Effect of DCT and PVD treatments on the fatigue behaviour of AISI 302 stainless steel. *Key Eng. Mater.* **2010**, *417*, 49–52. [CrossRef]

102. Baldissera, P.; Delprete, C. Effects of deep cryogenic treatment on static mechanical properties of 18NiCrMo5 carburized steel. *Mater. Des.* **2009**, *30*, 1435–1440. [CrossRef]

103. Baldissera, P. Fatigue scatter reduction through deep cryogenic treatment on the 18NiCrMo5 carburized steel. *Mater. Des.* **2009**, *30*, 3636–3642. [CrossRef]

104. Baldissera, P.; Delprete, C. The formal analogy between Tanaka-Mura and Weibull models for high-cycle fatigue. *Fatigue Fract. Eng. Mater. Struct.* **2012**, *35*, 114–121. [CrossRef]

105. Bouzada, F.; Cabeza, M.; Merino, P.; Trillo, S. Effect of Deep Cryogenic Treatment on the Microstructure of an Aerospace Aluminum Alloy. *Adv. Mater. Res.* **2012**, *445*, 965–970. [CrossRef]

106. Lei, X.; Deng, Y.; Yin, Z.; Xu, G. Tungsten Inert Gas and Friction Stir Welding Characteristics of 4-mm-Thick 2219-T87 Plates at Room Temperature and −196 °C. *J. Mater. Eng. Perform.* **2014**, *23*, 2149–2158. [CrossRef]

107. Vahdat, S.E.; Nategh, S.; Mirdamadi, S. Effect of microstructure parameters on tensile toughness of tool steel after deep cryogenic treatment. *Int. J. Precis. Eng. Man.* **2014**, *15*, 497–502. [CrossRef]

108. Niaki, K.S.; Vahdat, S.E. Fatigue Scatter of 1.2542 Tool Steel after Deep Cryogenic Treatment. *Mater. Today Proc.* **2015**, *2*, 1210–1215. [CrossRef]

109. Senthilkumar, D. Effect of deep cryogenic treatment on residual stress and mechanical behaviour of induction hardened En 8 steel. *Adv. Mater. Res.* **2016**, *2*, 427–436. [CrossRef]

110. Araghchi, M.; Mansouri, H.; Vafaei, R.; Guo, Y. A novel cryogenic treatment for reduction of residual stresses in 2024 aluminum alloy. *Mater. Sci. Eng. A* **2017**, *689*, 48–52. [CrossRef]

111. Gu, K.; Zhao, B.; Weng, Z.; Wang, K.; Cai, H.; Wang, J. Microstructure evolution in metastable β titanium alloy subjected to deep cryogenic treatment. *Mater. Sci. Eng. A* **2018**, *723*, 157–164. [CrossRef]

112. Nießen, F.; Villa, M.; Somers, M.A. Martensite formation from reverted austenite at sub-zero Celsius temperature. *Met. Mater. Trans. A* **2018**, *49*, 5241–5245. [CrossRef]

113. Kalsi, N.S.; Sehgal, R.; Sharma, V.S. Cryogenic treatment of tool materials: A review. *Mater. Manuf. Process.* **2010**, *25*, 1077–1100. [CrossRef]

114. Singla, A.K.; Singh, J.; Sharma, V.S. Processing of materials at cryogenic temperature and its implications in manufacturing: A review. *Mater. Manuf. Process.* **2018**, *33*, 1603–1640. [CrossRef]

115. Shokrani, A.; Dhokia, V.; Muñoz-Escalona, P.; Newman, S.T. State-of-the-art cryogenic machining and processing. *Int. J. Compd. Integr. Manuf.* **2013**, *26*, 616–648. [CrossRef]

116. Yong, A.Y.L.; Seah, K.H.W.; Rahman, M. Performance of cryogenically treated tungsten carbide tools in milling operations. *Int. J. Adv. Manuf. Technol.* **2007**, *32*, 638–643. [CrossRef]

117. Firouzdor, V.; Nejati, E.; Khomamizadeh, F. Effect of deep cryogenic treatment on wear resistance and tool life of M2 HSS drill. *J. Mater. Process. Technol.* **2008**, *206*, 467–472. [CrossRef]

118. Vadivel, K.; Rudramoorthy, R. Performance analysis of cryogenically treated coated carbide inserts. *Int. J. Adv. Manuf. Technol.* **2009**, *42*, 222–232. [CrossRef]

119. SreeramaReddy, T.V.; Sornakumar, T.; VenkataramaReddy, M.; Venkatram, R. Machinability of C45 steel with deep cryogenic treated tungsten carbide cutting tool inserts. *Int. J. Refract. Met. Hard Mater.* **2009**, *27*, 181–185. [CrossRef]

120. Reddy, T.S.; Sornakumar, T.; Reddy, M.V.; Venkatram, R.; Senthilkumar, A. Turning studies of deep cryogenic treated p-40 tungsten carbide cutting tool inserts—Technical communication. *Mach. Sci. Technol.* **2009**, *13*, 269–281. [CrossRef]

121. Yusof, N.M.; Razavykia, A.; Farahany, S.; Esmaeilzadeh, A. Effect of modifier elements on machinability of Al-20%Mg2Si metal matrix composite during dry turning. *Mach. Sci. Technol.* **2016**, *20*, 460–474. [CrossRef]

122. Sundaram, M.M.; Yildiz, Y.; Rajurkar, K.P. Experimental study of the effect of cryogenic treatment on the performance of electro discharge machining. In Proceedings of the ASME 2009 International Manufacturing Science and Engineering Conference, West Lafayette, ID, USA, 4–7 October 2009; pp. 215–222. [CrossRef]

123. Gill, S.S.; Singh, J. Effect of deep cryogenic treatment on machinability of titanium alloy (Ti-6246) in electric discharge drilling. *Mater. Manuf. Process.* **2010**, *25*, 378–385. [CrossRef]

124. Gill, S.S.; Singh, H.; Singh, R.; Singh, J. Flank Wear and Machining Performance of Cryogenically Treated Tungsten Carbide Inserts. *Mater. Manuf. Process.* **2011**, *26*, 1430–1441. [CrossRef]

125. Dogra, M.; Sharma, V.S.; Sachdeva, A.; Suri, N.M.; Dureja, J.S. Performance evaluation of CBN, coated carbide, cryogenically treated uncoated/coated carbide inserts in finish-turning of hardened steel. *Int. J. Adv. Manuf. Technol.* **2011**, *57*, 541–553. [CrossRef]

126. Shirbhate, A.D.; Deshpande, N.V.; Puri, Y.M. Effect of cryogenic treatment on cutting torque and surface finish in drilling operation with AISI M2 high speed steel. *Int. J. Mech. Eng. Rob. Res.* **2012**, *1*, 1–10. [CrossRef]

127. Çiçek, A.; Kıvak, T.; Uygur, I.; Ekici, E.; Turgut, Y. Performance of cryogenically treated M35 HSS drills in drilling of austenitic stainless steels. *Int. J. Adv. Manuf. Technol.* **2012**, *60*, 65–73. [CrossRef]

128. Kumar, A.; Maheshwari, S.; Sharma, C.; Beri, N. Machining efficiency evaluation of cryogenically treated copper electrode in additive mixed EDM. *Mater. Manuf. Process.* **2012**, *27*, 1051–1058. [CrossRef]

129. Iranmanesh, S.; Esmaeilzadeh, A.; Razavykia, A. Optimization of Electrical Discharge Machining Parameters of Co-Cr-Mo Using Central Composite Design. In Proceedings of the SDM 2017 Sustainable Design and Manufacturing, Bologna, Italy, 26–28 April 2017; pp. 48–57. [CrossRef]

130. Yuan, Y.G.; Xu, C.H. Influence of cryogenic treatment on grinding residual stress of WC-Co cemented carbides. *Adv. Mater. Res.* **2012**, *538*, 1746–1750. [CrossRef]

131. Çiçek, A.; Uygur, I.; Kıvak, T.; Özbek, N.A. Machinability of AISI 316 austenitic stainless steel with cryogenically treated M35 high-speed steel twist drills. *J. Manuf. Sci. Eng.* **2012**, *134*, 061003. [CrossRef]

132. Kapoor, J.; Singh, S.; Khamba, J.S. Effect of cryogenic treated brass wire electrode on material removal rate in wire electrical discharge machining. *Proc. Inst. Mech. Eng. Part C J. Mech. Eng. Sci.* **2012**, *226*, 2750–2758. [CrossRef]

133. Aoyama, S. High-performance coated wire electrodes for high-speed cutting and accurate machining. *Hitachi Cable Rev.* **1999**, *18*, 75–78.

134. Razavykia, A.; Yavari, M.R.; Iranmanesh, S.; Esmaeilzadeh, A. Effect of electrode material and electrical discharge machining parameters on machining of CO-CR-MO. *Int. J. Mech. Mechatr. Eng.* **2016**, *16*, 53–61.

135. Çiçek, A.; Kara, F.; Kivak, T.; Ekici, E. Evaluation of machinability of hardened and cryo-treated AISI H13 hot work tool steel with ceramic inserts. *Int. J. Refract. Metals Hard Mater.* **2013**, *41*, 461–469. [CrossRef]

136. Jafferson, J.M.; Hariharan, P. Machining performance of cryogenically treated electrodes in microelectric discharge machining: A comparative experimental study. *Mater. Manuf. Process.* **2013**, *28*, 397–402. [CrossRef]

137. Özbek, N.A.; Çiçek, A.; Gülesin, M.; Özbek, O. Investigation of the effects of cryogenic treatment applied at different holding times to cemented carbide inserts on tool wear. *Int. J. Mach. Tools Manuf.* **2014**, *86*, 34–43. [CrossRef]

138. Mavi, A.; Korkut, I. Machinability of a Ti-6Al-4V alloy with cryogenically treated cemented carbide tools. *Mater. Technol.* **2014**, *48*, 577–580.

139. Kumar, S.; Batish, A.; Singh, R.; Singh, T.P. A mathematical model to predict material removal rate during electric discharge machining of cryogenically treated titanium alloys. *Proc. Inst. Mech. Eng. Part B J. Eng. Manuf.* **2015**, *229*, 214–228. [CrossRef]

140. He, H.B.; Han, W.Q.; Li, H.Y.; Li, D.Y.; Yang, J.; Gu, T.; Deng, T. Effect of deep cryogenic treatment on machinability and wear mechanism of TiAlN coated tools during dry turning. *Int. J. Precis. Eng. Man.* **2014**, *15*, 655–660. [CrossRef]

141. Thornton, R.; Slatter, T.; Lewis, R. Effects of deep cryogenic treatment on the wear development of H13A tungsten carbide inserts when machining AISI 1045 steel. *Prod. Eng.* **2014**, *8*, 355–364. [CrossRef]

142. Thakur, D.G.; Ramamoorthy, B.; Vijayaraghavan, L. Effect of posttreatments on the performance of tungsten carbide (K20) tool while machining (turning) of Inconel 718. *Int. J. Adv. Manuf. Technol.* **2015**, *76*, 587–596. [CrossRef]

143. Dong, D.; Guo, G.; Yu, D.; An, Q.; Chen, M. Experimental investigation on the effects of different heat treatment processes on grinding machinability and surface integrity of 9Mn2V. *Int. J. Adv. Manuf. Technol.* **2015**, *81*, 1165–1174. [CrossRef]

144. Kıvak, T.; Şeker, U. Effect of cryogenic treatment applied to M42 HSS drills on the machinability of Ti-6Al-4V alloy. *Mater. Technol.* **2015**, *49*, 949–956. [CrossRef]

145. Özbek, N.A.; Çiçek, A.; Gülesin, M.; Özbek, O. Effect of cutting conditions on wear performance of cryogenically treated tungsten carbide inserts in dry turning of stainless steel. *Tribol. Int.* **2016**, *94*, 223–233. [CrossRef]

146. Xu, L.Y.; Zhu, J.; Jing, H.Y.; Zhao, L.; Lv, X.Q.; Han, Y.D. Effects of deep cryogenic treatment on the residual stress and mechanical properties of electron-beam-welded Ti-6Al-4V joints. *Mater. Sci. Eng. A* **2016**, *673*, 503–510. [CrossRef]

147. Khanna, R.; Singh, H. Comparison of optimized settings for cryogenic-treated and normal D-3 steel on WEDM using grey relational theory. *Proc. Inst. Mech. Eng. Part L J. Mater. Des. Appl.* **2016**, *230*, 219–232. [CrossRef]

148. Goyal, R.; Sngh, S.; Kumar, H. Performance evaluation of cryogenically assisted electric discharge machining (CEDM) process. *Mater. Manuf. Process.* **2018**, *33*, 433–443. [CrossRef]

149. Ghosh, S.; Rao, P.V. Performance evaluation of deep cryogenic processed carbide inserts during dry turning of Nimonic 90 aerospace grade alloy. *Tribol. Int.* **2017**, *115*, 397–408. [CrossRef]

150. Khan, A.; Maity, K. Comparative study of some machinability aspects in turning of pure titanium with untreated and cryogenically treated carbide inserts. *J. Manuf. Process.* **2017**, *28*, 272–284. [CrossRef]

151. Naveena, B.; Mariyam Thaslima, S.S.; Savitha, V.; Krishna, B.N.; Raj, D.S.; Karunamoorthy, L. Simplified MQL system for drilling AISI 304 SS using cryogenically treated drills. *Mater. Manuf. Process.* **2017**, *32*, 1679–1684. [CrossRef]

Review

Effect of Different Surface Conditions on Toughness of Vanadis 6 Cold Work Die Steel—A Review

Peter Jurči

Faculty of Materials Science and Technology in Trnava, Slovak University of Technology in Bratislava, Jána Bottu 2781/25, 917 24 Trnava, Slovakia; p.jurci@seznam.cz

Received: 7 May 2019; Accepted: 20 May 2019; Published: 22 May 2019

Abstract: The effects of surface roughness, presence of nitrided diffusion regions, and magnetron sputtering of Cr_2N–6Ag thin films on the toughness of Cr–V ledeburitic Vanadis 6 die steel were investigated by using the flexural strength measurement method, which was coupled with careful microstructural investigations and analyses of fractured surfaces. The results undoubtedly show that enhanced surface roughness reduces the material toughness, since the cusps formed on the metallic surface as a result of the machining act as preferential sites for crack nucleation and growth. The presence of nitrided regions on the surface, on the other hand, forms a structural notch there, which has a strong detrimental effect on toughness. Deposition of Cr_2N–6Ag thin films has only marginal effect on the steel toughness. Practical recommendations for the designers, heat treaters, and coaters of the tools are thus that they should maintain the surface finish quality of the tools as high as possible, avoid too thick and supersaturated nitrided regions, and that there is almost no risk of tool embrittlement due to physical vapor deposition (PVD) coating.

Keywords: Vanadis 6 die steel; surface finish; nitriding; PVD coating; toughness; fractography

1. Introduction

High-carbon, high-chromium, and high-vanadium cold work tool steels have been frequently used in modern industries in applications where superior wear resistance and strength are required. This includes industrial branches like powder compacting, paper cutting, sheet forming, and fine blanking.

The high strength and hardness, and excellent wear resistance of these tool steel grades are a result of the standard heat treatment procedure. It comprises vacuum austenitizing, holding at the desired temperature for a pre-determined duration, and inert gas quenching, followed by immediate tempering. Properly performed heat treatment provides the steels with a hardness of around 60 HRC at the secondary hardness peak. Excellent wear resistance is ensured by a high amount of carbides, among which the MC particles play a dominant role.

However, ledeburitic steels generally have low toughness. Moreover, these materials suffer from poor and inhomogeneous carbide distribution when manufactured by classical ingot metallurgy. This causes an "extra" embrittlement of the materials and a great level of anisotropy of the mechanical properties [1]. Advanced powder metallurgy (PM) ledeburitic steels do not manifest anisotropy of the mechanical properties [2]. Despite that, their toughness (being represented by flexural strength) is limited by approx. 4000 MPa, and the fracture toughness is around 15–16 $MPa \cdot m^{1/2}$ [3], at a hardness of 60–61 HRC.

After the heat treatment, the tools should be surface finished before their use. Surface finishing involves machining operations like fine grinding, lapping, and polishing. These operations provide the tools with the final surface roughness, which is specific for each of them. Fine grinding, for instance, results in surface roughness (R_a) of 0.2–0.4 µm, while lapping reduces the R_a to a level of 0.02 µm, and polishing provides the metallic surface with a so-called "mirror finish".

It is well known that the surface roughness may have important impacts on the toughness of brittle materials like dental ceramics, as pointed out by Vasconcellos Amarante et al. [4] and by Hallmann et al. [5]. One can expect that the toughness of heat-treated ledeburitic tool steels will also be influenced by surface roughness, as they belong to the group of relatively brittle materials. Despite that, the information on the effect of surface roughness on fracture performance of ledeburitic steels is lacking, and there is no comprehensive study on this effect published yet. Only Spies, Riese, and Hoffmann [6] reported that the flexural strength of hardened and tempered M2 high-speed steel decreased from 3780 to 3700 and 3600 MPa when the surface roughness increased from 0.1 to 0.5 and 1 μm, respectively.

Plasma nitriding has been widely used in steels, resulting in benefits such as the increase of fatigue life [7], the improvement of tribological properties [8,9], and the increase of corrosion resistance [10]. The plasma nitriding applied to tool steels has recently gained a new impulse with the development of duplex coating, which is a combined treatment consisting of the deposition of a hard physical vapor deposition (PVD) layer on a pre-nitrided surface [11–16]. The presence of an intermediate hardened layer under very hard thin films has shown benefits for the mechanical properties of the surface. Bell et al. [17], for instance, suggested a better load bearing capacity, improved fatigue resistance, and a smooth hardness profile from the surface to the substrate, resulting in lower levels of residual stresses in the layer/substrate interface. It was also demonstrated that the duplex coating leads to much better adhesion of thin films to the substrates [12,13,16] and higher wear resistance [12,14], and thereby leads to manyfold extended tool service life [11,15,18].

All the cited works show the benefits of nitriding the surfaces of tool steels. Nevertheless, nitrogen input to the surface makes the steel harder and therefore more brittle. Embrittlement of ledeburitic tool steels has been indicated by Hock et al. [19], and later experimentally proved by Kwietniewski et al. [20] by industrial testing of nitrided and duplex-coated single-point turning tools. Unfortunately, the embrittlement has not been determined exactly but only estimated based on evaluation of tools damage as well as by microstructural evaluation of nitrided surfaces. Hence, an exact assessment of the effect of a nitrided surface on the bulk toughness is not available in the scientific literature to date.

Thin ceramic films based on transition metal nitrides (carbides, oxides) have been proven to be good candidates for protecting the base tool steel against wear, corrosion, and other unwanted damage.

In the field of thin solid films, PVD (physical vapor deposition)-produced titanium nitride (TiN) is still the most widely accepted in engineering applications. The combination of high hardness, wear resistance, chemical inertness, and low friction coefficient characteristics in TiN makes it attractive as a tribological coating. However, the main drawback of TiN is its limited oxidation resistance (approximately 500 °C). The resistance against high-temperature oxidation can be improved by the addition of Al and Cr [21,22]. TiAlN coatings have been developed as alternatives to TiN because of their higher corrosion resistance (up to 750 °C) and higher hardness [23]. But their wear performance at ambient temperature is negatively influenced due to their high friction coefficient [24].

Another alternative to TiN are the Cr_xN_y films, which have been developed over the past three decades. They have gained early popularity in a variety of industrial applications due to superior wear resistance [25,26], good corrosion resistance [27–29], and good cutting properties in copper machining [30], aluminium die casting [29], or in woodworking [31,32]. Cr_xN_y films can be synthesized in wide ranges of chemistries, phase constitutions, and properties. The microhardness of Cr_xN_y films ranges between 1500 and 3000 HV [30,33–36]. The Young modulus can also be varied, from the lowest values of around 188 to the highest ones of 315 GPa [33,35]. The adhesion of Cr_xN_y thin films can be influenced by their phase constitution [37] and internal stress level [37], but also by proper selection of the substrate. Adhesion is better when hard substrates (ledeburitic steels, cemented carbides) are coated [32,34,38]. Later, it was found that small additions of silver can provide the Cr_xN_y films with self-lubricating properties and very low friction coefficients, with almost unaffected or rather slightly improved nanohardness and adhesion [39–42]. Despite the high universality of Cr_xN_y films, their use is not possible in selected cases. Carlsson and Olsson [43], for instance, concluded that metal–carbide-doped diamond-like carbon (DLC) coatings are good candidates for tool coatings in the dry cold forming of hot dip Zn steel sheets. These steel sheets are not suited for dry forming due to severe material pick-up, and the use of Cr_xN_y coating fails in these cases due to high friction coefficients and material transfer from the steel sheet surfaces.

Even though a great number of scientific papers have been published about the growth, microstructure, and key properties of PVD thin films, only a little attention has been paid to the effect of the presence of thin ceramic films on the substrate toughness.

To the best knowledge of the author, no comprehensive and overview report has been made so far to review the influence of different surface treatments on the toughness of ledeburitic steels. The major aim of the present overview is to overcome this limitation and to summarize, review, and discuss the existing information about the effect of different surface states on the toughness of powder metallurgy cold work die steel Vanadis 6. The impacts of different surface roughness developed by machining operations, presence of plasma-nitrided regions with different thickness and microhardness, and influence of CrAgN nanocomposite thin films on the flexural strength (toughness measure) of the experimental material will be presented, analyzed, and discussed. In the text, the original papers will be referred to in some places in order to avoid potential confusions.

2. Description of Experimental Material, Processing, and Investigations Methods

A commercially available PM ledeburitic cold work tool steel Vanadis 6 (Uddeholm AB, Hagfors, Sweden) with a nominal composition (wt.%) of 2.1% C, 1.0% Si, 0.4% Mn, 6.8% Cr, 1.5% Mo, 5.4% V, and Fe as the balance was used as an experimental material. The initial state of the material was soft-annealed, with a hardness of 284 HV10. The microstructure of the steel manifests a high degree of isotropy. Hence, the orientation of the semi-finished product was disregarded in sampling [2].

Two types of specimens were manufactured. The first ones were cylinders with 20 mm in diameter and 6 mm in thickness for microstructural examinations, and the second group of specimens was plates with dimensions (thickness × width × length) of 1 × 10 × 100 mm (Figure 1) for flexural strength determination. All the specimens were milled to a surface roughness of 6 μm and subjected to the pre-determined heat treatment schedule. The heat treatment consisted of the following steps: Gradual heating up to the desired austenitizing temperature T_A (1050 °C) in a vacuum furnace and then holding at that temperature for 30 min to homogenize the austenite, which was followed by quenching by nitrogen gas (5 bar pressure). Immediately after quenching to the room temperature, the specimens were moved to a tempering furnace. Double tempering was carried out at a temperature of 530 °C, with a duration of each tempering cycle of 2 h. The material was cooled down slowly to the room temperature after each tempering cycle.

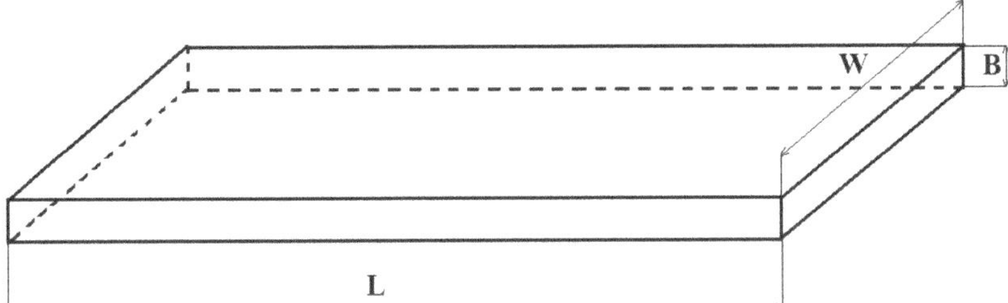

Figure 1. A schematic of the specimens used for flexural strength determination.

Then, the specimens were divided to several batches (Table 1).

Table 1. Characterization of specimen batches used in the experimental works.

Batch	Surface Finish	Post Heat Treatment	Total Number of Specimens
1	Milled, $R_a = 6\ \mu m$	No	5
2	Ground, $R_a = 0.3\ \mu m$	No	5
3	Mirror finish polished, $R_a = 0.017\ \mu m$	No	5
4	Mirror finish polished, $R_a = 0.017\ \mu m$	Plasma nitriding	15
5	Mirror finish polished, $R_a = 0.017\ \mu m$	Coating with Cr_2N-6Ag	5

Batches 1, 2, and 3 were examined without any post-heat treatment. Batch 4 was subjected to the plasma nitriding, and the specimens from Batch 5 were coated with a Cr_2N–6Ag nanocomposite thin film.

Plasma nitriding was performed in a RUBIG PN 60/60 (Rubig GmBH, Wels, Austria) device at different temperatures and for various processing durations (Table 2). Five specimens (from Batch 4; Table 1) were treated at each of processing parameter combinations. First, the specimens were cleaned and degreased in an ultrasonic acetone bath for 15 min. Then, they were moved to the plasma nitriding device, where they were heated up to the pre-determined temperature, and sputter cleaned for 30 min in a pure hydrogen atmosphere in order to activate the surface. Afterwards, an atmosphere containing nitrogen and hydrogen in a ratio of 1:3, at a pressure of 300 Pa, was introduced into the processing chamber. The plasma nitriding procedures were carried out by using a voltage of 500 V, in a pulse regime with the pulse time of 100 μs. After the plasma nitriding, the specimens were cooled down slowly to the room temperature.

Table 2. List of plasma nitriding parameters applied for the treatment of specimens.

Batch	Temperature (°C)	Duration (min)
4.1	470	30
4.2	500	60
4.3	530	120

Cr_2N–6Ag thin films were synthesized in a Hauzer-Flexicoat 850 (Hauzer, Venlo, The Netherlands) magnetron sputter deposition system, in a pulse regime with a frequency of 40 kHz. For the growth of the Cr_2N–6Ag films, two water-cooled cathodes, the first one made of chromium with the standard purity of 99% and the second one manufactured out of silver with the purity of 99.98%, were used. The cathodes were positioned opposite to one another, at an angle of 45° with respect to the substrate surface. The output power of the Cr cathode was 5.8 kW, and that of the Ag cathode was 0.2 kW, in order to add 6 mass% of Ag into the base Cr_2N.

The substrates were placed between the targets on rotating holders with a rotation speed of 3 rpm. At the beginning of the process, pure Ar (99.999% purity) was introduced into the processing chamber in order to sputter clean the substrates. During the cleaning step, the substrate temperature reached 250 °C, the negative substrate bias was 200 V, and the duration of sputter cleaning was 15 min. Just prior to deposition, both targets were sputter cleaned for 7 min, with the samples covered with a protective shutter during this period. Next, pure N_2 (99.998% purity) was introduced into the chamber to obtain an atmospheric composition of N_2:Ar = 1:4.5 and a pressure of 0.15 mbar. A negative substrate bias of 100 V was used for the growth of the films. The deposition temperature was increased to 500 °C by the use of internal wall resistive heaters.

The microstructure of the examined steel was evaluated by light microscopy, by using a Neophot 32 apparatus (Carl Zeiss AG, Oberkochen, Germany). The same light microscope was used for the inspection of the surface morphologies of the specimens with different roughness as well as for microstructural investigations of differently plasma nitrided specimens. The specimens were Nital (3% ethanol solution of nitric acid) etched after standard metallographic preparation.

For more precise examinations, however, scanning electron microscopy (SEM) and transmission electron microscopy (TEM) were used. A JEOL JSM 7600 F (Jeol Ltd., Tokyo, Japan) apparatus equipped with an energy-dispersive spectroscopy (EDS) detector (Oxford Instruments, plc., High Wycombe,

UK) and a wavelength-dispersive spectrometry (WDS) detector was used for SEM observations. The microstructure was recorded in the secondary electron (SE) detection regime. The microstructure of magnetron-sputtered thin films was examined on the fractured surfaces of specimens prepared as follows: After deposition, the specimens were immersed into liquid nitrogen, held there for 15 min, and broken down. Then, they were reheated slowly to room temperature, cleaned ultrasonically in acetone, and dried before inserting into the SEM. TEM was carried out by using a JEOL 200CX microscope (Jeol Ltd., Tokyo, Japan) operating at an acceleration voltage of 200 kV. Thin foils were made by a standard preparation technique. Pieces with square cross-sections and thickness of 0.15 mm were cut off from the material, and mechanically thinned to a thickness of approximately 20 μm. The final thinning was realized by using an electrolytic jet-polisher (Tenupol 5).

Microhardness depth profiles throughout the nitrided regions and the base material microhardness were measured with a Buehler Indentament 1100 tester (Buehler Ltd., Lake Bluff, IL, USA), at a load of 50 g (HV 0.05) and loading time of 20 s.

The nanohardness and the Young´s modulus (E) of Cr_2N–6Ag thin films were measured by using a NanoTest (Micro Materials Ltd, Wrexham, UK) nanohardness tester. The maximum normal load was 20 mN, and a Berkovich indenter was employed. Ten measurements were made, and the mean value and the standard deviation were then calculated. The penetration depth (and loading) was chosen to not exceed one tenth of the total coating thickness.

In order to analyze the chemical composition throughout the Cr_2N–6Ag thin films, Glow Discharge Optical Emission Spectroscopy (GDOES) by using a Spectruma GDA 750 (Spectruma Analytik GmbH, Hof, Germany) device was carried out.

Flexural strength was determined by using an Instron 8862 testing device (Instron, Norwood, MA, USA), with the distance between loading roller supports of 80 mm, at an ambient temperature. Specimens were loaded in three-point bending at a loading rate of 1 mm/min, up to the moment of fracture. Flexural strength was determined from the maximum (fracture) load according to a standard approach. In addition, the total work of fracture (W_{of}) was determined as deformation energy evaluated from corresponding area below the measured load–deflection (load–displacement) curve.

The flexural strength was calculated using the formula:

$$R_{mo} = \frac{3 \times F \times l_0}{2 \times B^2 \times W} \tag{1}$$

where R_{mo} is the flexural strength, F is the load at the fracture point, L_o is the distance between supports, and B and W are the thickness and width of the specimen, respectively.

The fracture surface morphology was investigated by a JEOL JSM 7600F scanning electron microscope.

3. Results, Discussion, and Practical Recommendations

3.1. Microstructure of Examined Steel

Figure 2 is a compilation of differently acquired micrographs showing the microstructure of the examined steel after realization of the heat treatment schedule. An optical micrograph, shown in Figure 2a, demonstrates that the microstructure of the steel is composed of tempered martensite and undissolved carbides, which are uniformly distributed throughout the martensitic matrix. The sizes of the carbides range from several hundreds of nm up to approx. 3 μm. A more detailed SEM micrograph, shown in Figure 2b, enables us to distinguish between eutectic carbides (ECs), secondary carbides (SCs), and small globular carbides (SGCs), according to the classification published recently [44]. The martensite manifests clear lath morphology, with no presence of retained austenite as this phase is decomposed by high temperature tempering. A bright-field TEM micrograph, shown in Figure 2c, shows that many of the martensitic laths are internally twinned. The widths of the twins are of around 20–30 nm. There are a lot of precipitates located inside the twins. This is sufficient for obtaining the dark-field image, shown in Figure 2d, as well as for obtaining a good intensity of reflections in the corresponding diffraction patterns, as shown in Figure 2e. Analysis of the reflections disclosed that the particles are of cementitic nature.

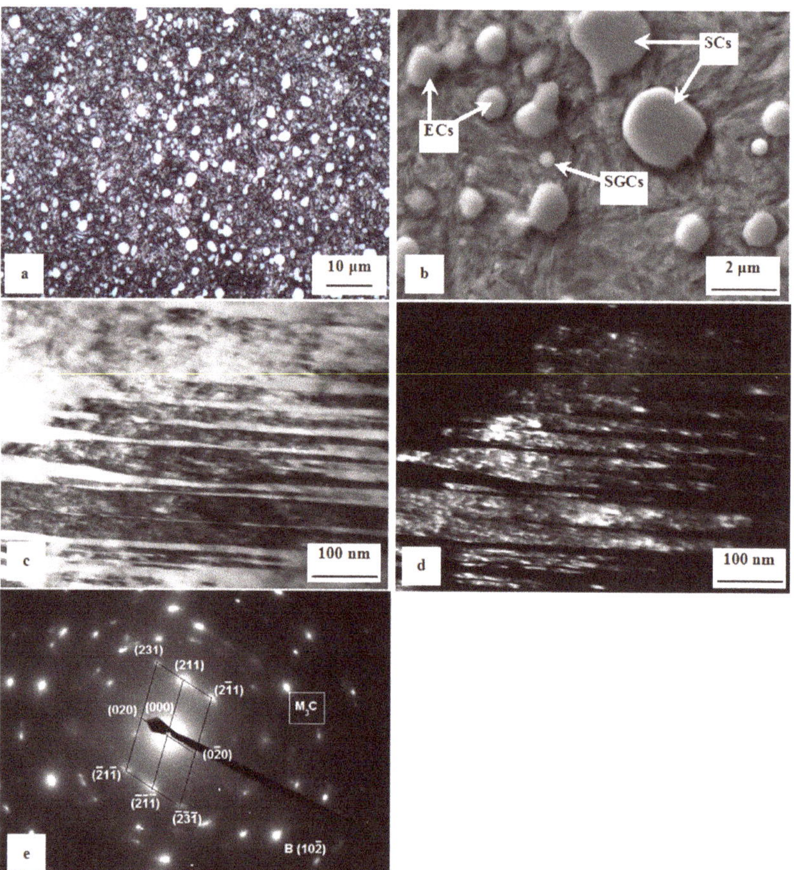

Figure 2. Microstructure of the experimental material after the given heat treatment schedule. (**a**) Optical micrograph, (**b**) SEM micrograph, (**c**) bright-field TEM micrograph showing internally twinned martensitic laths, (**d**) corresponding dark-field micrograph showing precipitates inside the martensitic domain, and (**e**) diffraction patterns of the precipitates.

The as-tempered hardness of the material was 733 ± 9 HV10, and the fracture toughness was determined to be 15.72 ± 0.31 MPa·m$^{1/2}$ [3].

3.2. Effect of Surface Roughness

The effect of surface roughness on toughness of the Vanadis 6 steel was investigated in our recent work [45]. Figure 3 shows plan-view optical micrographs showing the surface finishes of the tested specimens. After milling, the surface roughness was 6 µm (Figure 3a). Subsequent grinding led to a substantial reduction of surface roughness, to a value of 0.3 µm (Figure 3b). The third batch of the specimens was subjected to mirror polishing, and the surface was then very smooth (Figure 3c).

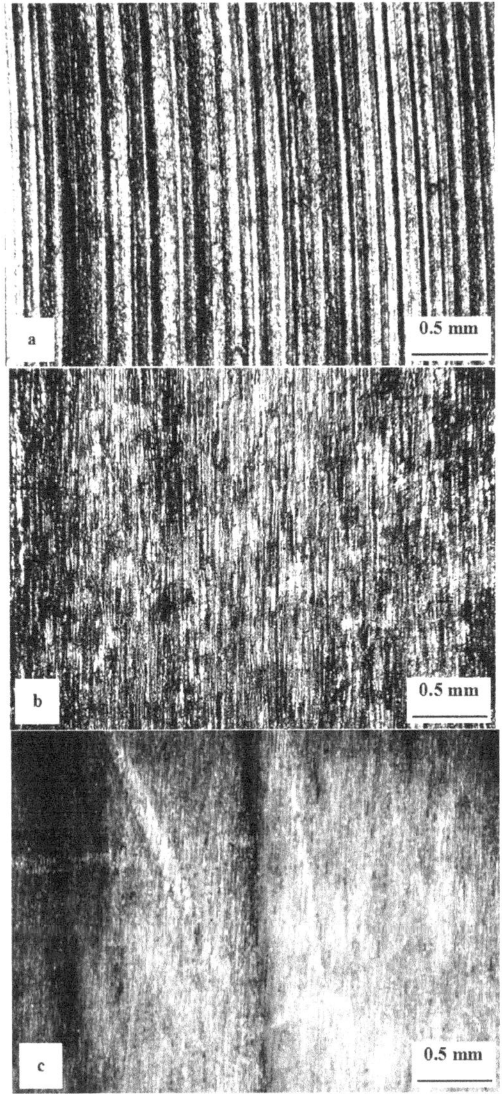

Figure 3. Plan-view optical micrograph showing the surfaces of specimens made of Vanadis 6 steel with different roughness: (**a**) Milled, (**b**) fine ground, and (**c**) polished. Adapted from the corresponding literature [45]. (Copyright, Materials Engineering/Materiálové inžinierstvo, 2011).

The load–flexure diagrams obtained by the flexural strength testing of specimens with different surface roughness are shown in Figure 4. It is shown that all the curves manifest the load–flexure dependence typical for hard and relatively brittle steels, i.e., there is linear dependence of these two quantities evident up to relatively high loading, which is followed by a small region suggesting limited plastic behavior of the material at higher loading. Alternatively, there is also an apparent difference between the load–flexure traces visible on the diagram—the load at the fracture moment decreases with increasing surface roughness.

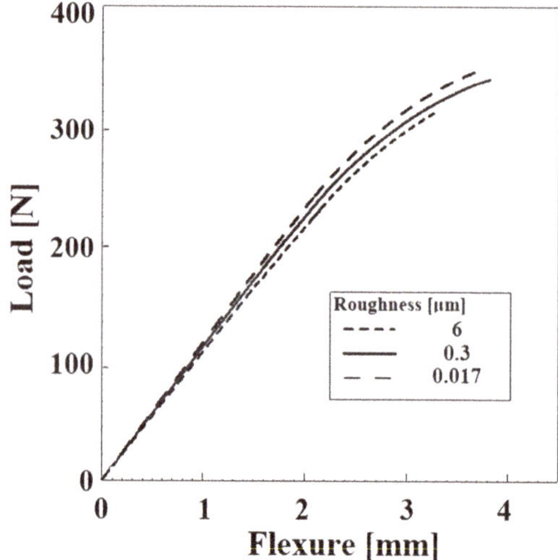

Figure 4. Load–flexure diagrams of the specimens with different surface roughness.

Figure 5 shows the dependence of flexural strength on the surface finish. It is shown that the flexural strengths for both the polished specimens (R_a = 0.017 μm) and the fine ground ones (R_a = 0.3 μm) were of around 4000 MPa, and the differences lie within the range of statistical uncertainty of the obtained results, even though the polished surface state seems to be more convenient from the point of view of high-material toughness maintenance (due to its slightly higher mean value as well as smaller scatter of the obtained results). Alternatively, the flexural strength of the milled material (R_a = 6 μm) manifests a clear decrease.

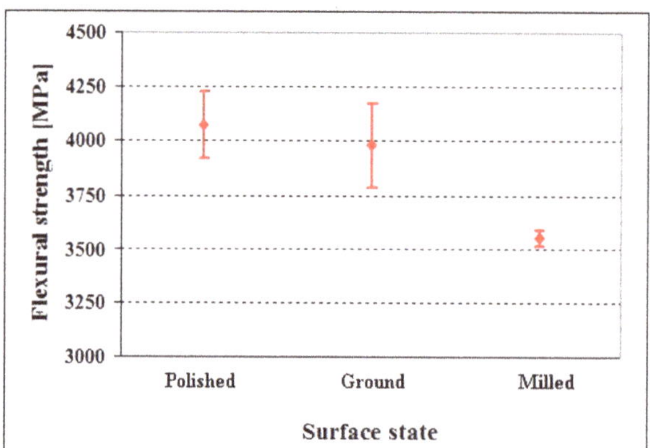

Figure 5. Flexural strengths of specimens made of the Vanadis 6 steel with different surface finish. Adapted from the corresponding literature [45]. (Copyright, Materials Engineering/Materiálové inžinierstvo, 2011).

The differences in flexural strength values, shown in Figure 5, are reflected in the values of the work of fracture, shown in Figure 6. As shown here, the work of fracture of polished material is almost the same as that of ground steel, and the differences lie within the range of statistical uncertainty of the results. On the other hand, the work of fracture of milled steel is lower despite its statistical uncertainty range considerably overlapping with those calculated for both the polished and ground specimens.

In other words, the dependence of the work of fracture on the surface finish closely follows the flexural strength dependence on the same surface parameter.

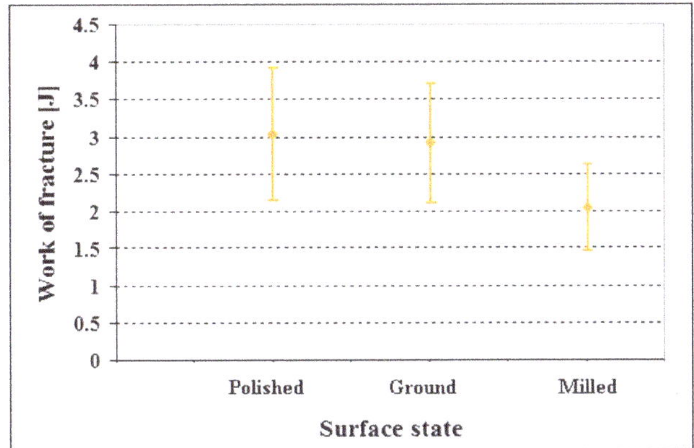

Figure 6. Work of fracture of specimens made of the Vanadis 6 steel with different surface finish. Adapted from the corresponding literature [45]. (Copyright, Materials Engineering/Materiálové inžinierstvo, 2011).

A compilation of SEM micrographs showing fractured surfaces of specimens with different surface roughness is shown in Figure 7. All the fractured surfaces manifest combined low-energetic ductile and cleavage morphology of the surface. The ductile component of the fracture is initiated mainly at the carbide/matrix interfaces, where microvoids are formed as a result of different plasticities (and stiffness) of the matrix and carbides. The carbides that assist in decohesive crack propagation (note that the formation of microvoids results from the decohesion at the phase interfaces) are denoted as DCs (Figure 7a,c,e). Coarser carbide particles undergo cleavage mechanisms of crack propagation more easily, and these carbides are marked as FCs (fractured carbides). The matrix mainly manifests so-called "dimple" morphology of fractures, which is associated with micro-plastic deformation (MPD), mainly on the sites of carbide/matrix interfaces. However, cleavage facettes (CLF) are also visible on the micrographs (Figure 7b,d,f). Even though the fractured surfaces appear very similar with respect to their morphology, the dimples at the carbide/matrix interfaces are deeper in the case of the specimens with better surface quality (polished). A comparison is shown in Figure 7b,d,f. This is due to higher levels of plastic deformation before the fracture, and is in good agreement with the results of three-point bending strength measurements (Figure 5), and also with the determination of the deformation energy (work of fracture), shown in Figure 6. These results suggest unambiguously that the surface quality plays an important role in the fracture behavior of relatively brittle materials like quenched and tempered PM ledeburitic steels.

These findings have great importance for the end-users of materials and tools, recommending them to make the surface as high quality as possible to prevent the initiation of cracks and to ensure the service reliability of tools. However, it should also be mentioned that the surface quality finish of tools can be reflected in the resulting surface roughness of workpieces made of different materials, which in turn influences their mechanical properties and durability. For instance, Bayoumi and Abdellatif [46], Javidi et al. [47], Sasahara [48], Noll and Erickson [49], and Taylor and Glancy [50] established that the fatigue strength reduces with increasing surface roughness in the cases of aluminium alloys, nickel–molybdenum alloys, and carbon steels, due to the stress concentrations generated by rough surfaces. Novovič et al. [51] stated that surface roughness values over 0.1 μm deteriorate the fatigue life on any component significantly. One can thus conclude that the tools' surface finish quality not only influences the toughness characteristics of the tools themselves, but also the overall production quality.

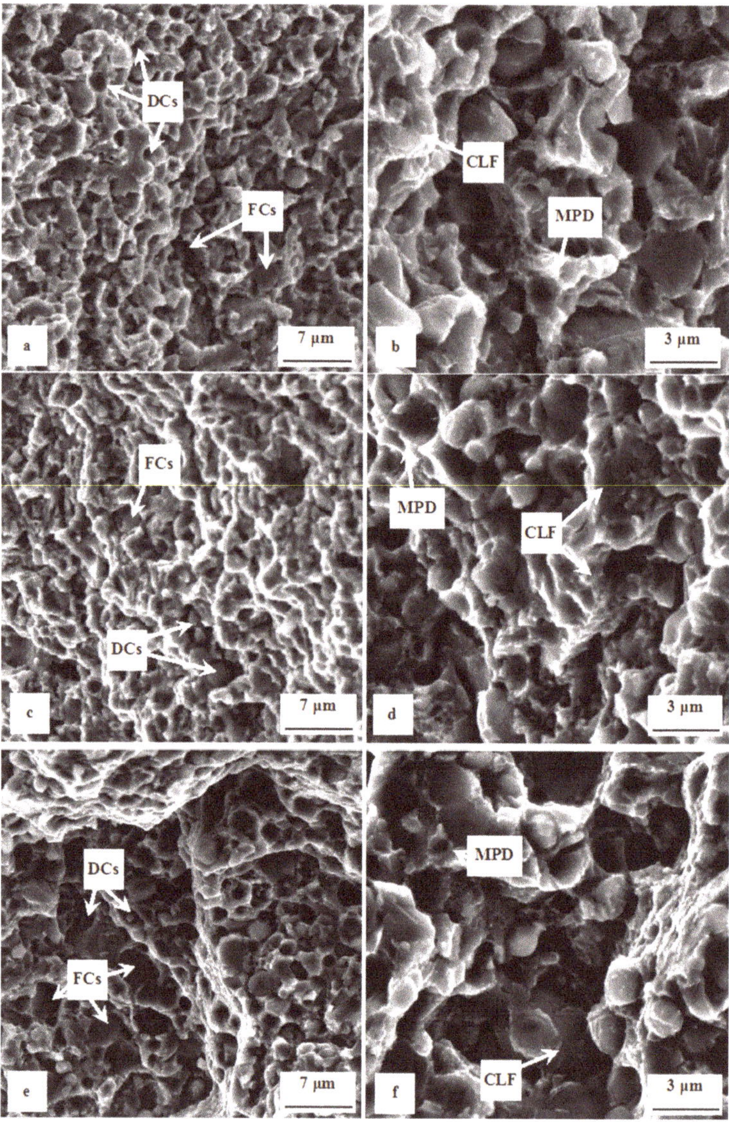

Figure 7. SEM micrographs showing fracture surfaces of specimens made of Vanadis 6 steel with different surface quality. (**a**,**b**) Milled (overview, detail); (**c**,**d**) fine ground (overview, detail); and (**e**,**f**) polished (overview, detail). MPD—micro-plastic deformation of the matrix, CLF—cleavage fracture, FCs—fractured carbides, DCs—carbides that assist in decohesive fracture mechanisms (decohesive carbides).

3.3. Effect of Plasma Nitriding

Figure 8 is a compilation of optical micrographs showing the microstructure of plasma nitrided regions developed on the surfaces of Vanadis 6 steel, by application of various treatment regimes. In Figure 8a, there is a nitrided region developed at the temperature of 470 °C for the duration of 30 min. The nitrided region differs from the substrate in terms of etching sensitivity, which is caused by the precipitation of nitrides. However, the nitrogen input into the material is relatively low in this case, which is reflected in only small differences in etching sensitivity between the nitrided surface and the substrate detectable in the optical micrograph. Nitriding at the temperature of 500 °C for 60 min highlights the differences in etching intensity between the nitrided region and bulk material. This is due to a higher input of nitrogen atoms into the steel substrate. However, neither the processing at 470 °C nor that at 500 °C leads to the formation of a compound "white" layer on the surface. Furthermore, the presence of nitride networks at the primary grain boundaries was not detected (Figure 8b). The micrograph in Figure 8c demonstrates that the nitriding at the temperature of 530 °C for the duration of 120 min leads to both the growth of a thin compound "white" layer on the surface (in this particular case, the white layer had a thickness of 1–1.5 μm) and formation of nitride networks at the original grain boundaries.

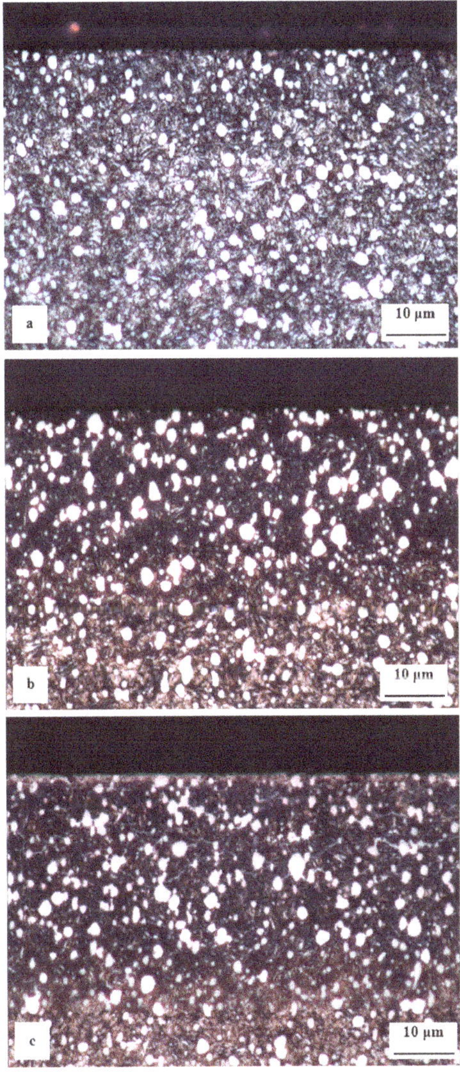

Figure 8. Optical micrographs showing the microstructures of differently plasma nitrided specimens: (**a**) Nitrided at 470 °C for 30 min, (**b**) nitrided at 500 °C for 60 min, and (**c**) nitrided at 530 °C for 120 min.

Table 3 summarizes the main parameters of nitrided regions. The surface concentration of nitrogen is 4.03 wt.% in the specimen nitrided at 470 °C for 30 min, and it increases moderately as the nitriding temperature is increased (or for longer processing durations). The nitrogen diffusion depths are 13.2, 42, and 67 μm for the specimens nitrided at 470, 500, and 530 °C, respectively, and these values correspond well with the values of nitriding case depth (Nht). The surface hardness values are 1316, 1564, and 1648 HV0.05 for the same specimens. In Figure 9, there are microhardness depth profiles for all these specimens. It is shown that the microhardness of the material nitrided at 470 °C for 30 min decreases rapidly with the increasing depth below the surface, while the microhardness depth profiles of other specimens appear relatively flat, with moderate decreases in the values.

Table 3. Main parameters of nitrided regions (*the nitriding case depth Nht is determined as core hardness + 50 HV0.05).

Nitriding Parameters	Surface Nitrogen Content (wt.%)	Nitrogen Diffusion Depth (μm)	Surface Microhardness (HV 0.05)	Nht* (μm)
470 °C/30 min	4.03	13.2	1316	13
500 °C/60 min	4.8	42	1564	39
530 °C/120 min	5.78	67	1648	60

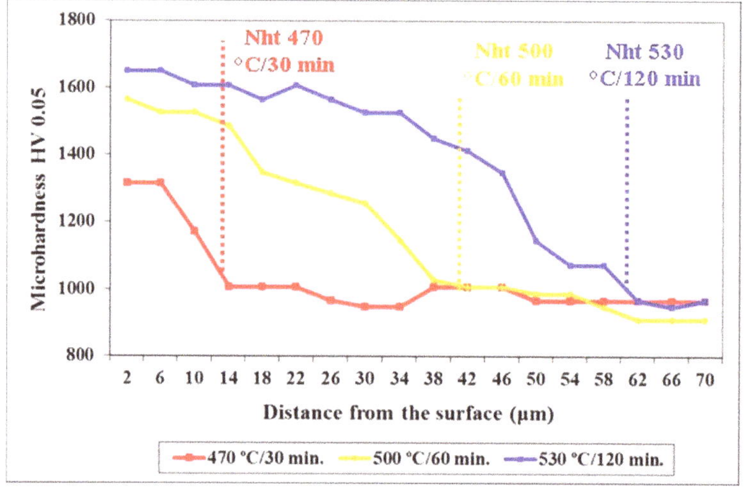

Figure 9. Microhardness depth profiles for differently nitrided specimens.

Figure 10 shows the flexural strengths of the materials as a function of plasma nitriding parameters. Non-nitrided material has the flexural strength of 4071 ± 154 MPa. The specimens nitrided at 470 °C for 30 min, 500 °C for 60 min, and 530 °C for 120 min have values of flexural strength of 2744 ± 282, 2414 ± 126, and 1861 ± 99 MPa, respectively. These results suggest that the diffusion regions on the surfaces reduce the flexural strength (and toughness) of the examined steel dramatically, and that this reduction increases with increasing the thickness of the nitrided region, its increasing saturation with nitrogen, hardness, etc.

The work of fracture of the non-nitrided specimen was 3.02 ± 0.88 J (Figure 11). Considerable reduction of flexural strength (Figure 8) is reflected in reduction of work of fracture. The values of W_{of} were 2.03 ± 0.49, 1.79 ± 0.44, and 1.38 ± 0.47 J for the differently plasma nitrided specimens.

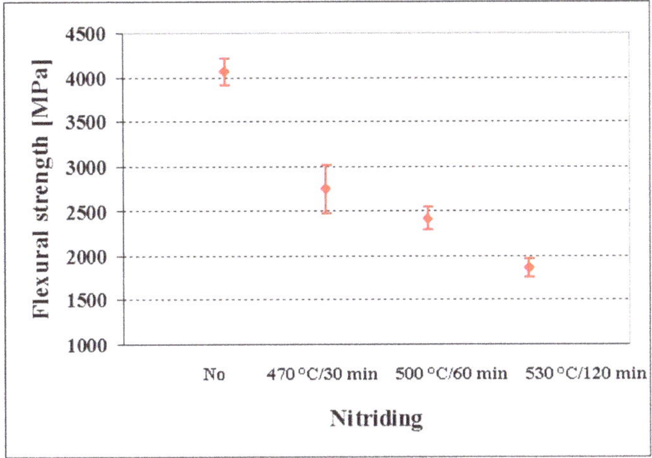

Figure 10. Flexural strength of non-nitrided and differently plasma nitrided specimens made of Vanadis 6 steel.

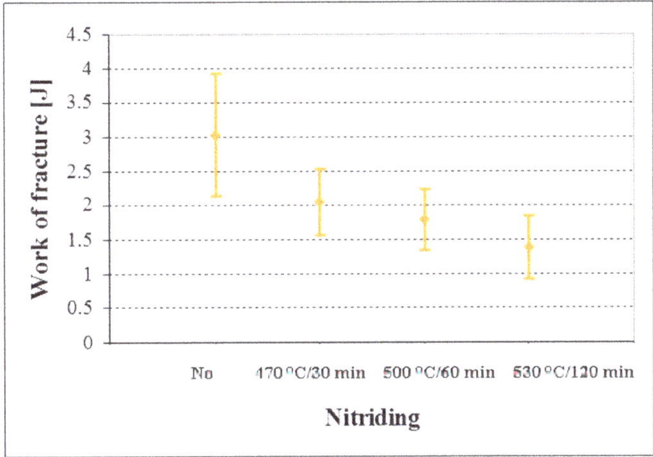

Figure 11. Work of fracture of non-nitrided and differently plasma nitrided specimens made of Vanadis 6 steel.

Furthermore, the fracture surfaces of nitrided steel reflect clearly the reduction of flexural strength. The surfaces manifest cleavage fractures (Figure 12) and the thickness of the cleavage regions corresponds well with the thickness of the nitrided regions. The fracture propagation manner through the bulk material is influenced by the presence of the nitrided region, and the topography of the fracture is significantly reduced (compare Figure 12, which is a detailed micrograph in a blue border, with Figure 7). The cleavage propagation manner, with no indication of plastic deformation, is a typical feature of nitrided regions, and the morphology of the fractured surface does not manifest any changes in different depths below the surface (Figure 12).

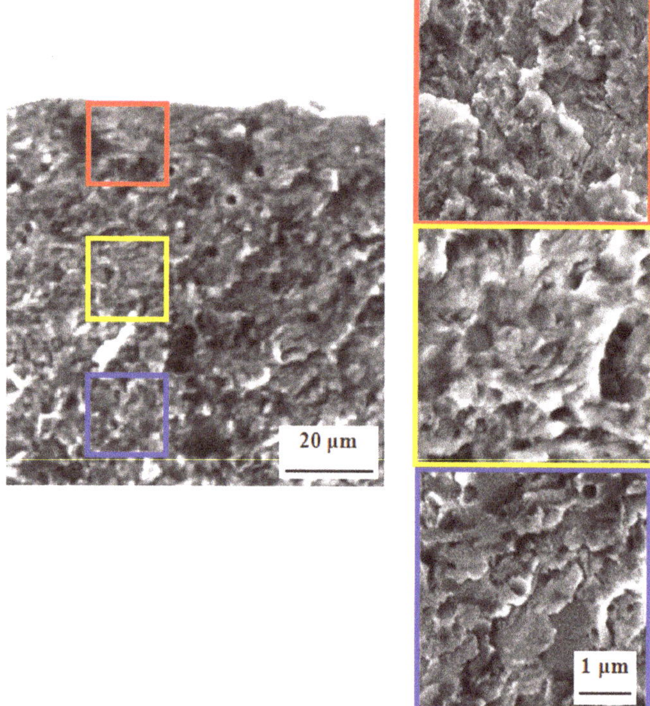

Figure 12. SEM micrographs showing fracture surface of specimen made of Vanadis 6 steel nitrided at 530 °C for 120 min.

The obtained results have great importance towards the industrial practice. Diffusion processes like nitriding or boronizing are widely used in industry in order to obtain benefits in hardness [7,52,53], wear performance [8,9], and corrosion behavior [10]. Furthermore, the industrial PVD coating companies often use nitriding as a pre-treatment prior to the deposition of hard ceramic layers, forming so-called duplex-coatings on the surface [11–17,54]. The duplex system (nitriding + PVD coating) brings many benefits to the surface treatments technology due to better support for the coating [16,54], which leads to much better adhesion of thin films to the steels [12,13,16]. However, the heat treaters must not forget the detrimental effects of the presence of nitrided layers on the toughness. In the work [55], it has been demonstrated that the flexural strength was significantly reduced not only for thin specimens, but that this reduction is practically independent of the specimen cross-section size. Hence, the nitriding processes should be controlled carefully in order to avoid the formation of either too thick nitrided regions or continuous networks of nitrides along the grain boundaries, and to maintain at least acceptable material toughness.

3.4. Effect of PVD Coating with Cr_2N-6Ag

The thickness of Cr_2N film with an addition of 6 mass% Ag is 4.3 µm, Figure 13. The film grew in a columnar manner, and the individual crystallites are well visible in the SEM micrograph. This growth manner is typical for magnetron-sputtered Cr_2N films developed at the same or similar combinations as the processing parameters [56–58]. Furthermore, it is shown in Figure 13a that silver forms individual agglomerates within the Cr_2N. The reason for this is that the silver is completely insoluble in CrN, as reported recently [39–42]. The plan-view SEM micrograph (Figure 13b) illustrates that the sizes of agglomerates range between several nm and 60 nm, and that the coarser Ag particles are located mainly between the individual Cr_2N crystallites, while the finer ones can also be found within the crystallites of Cr_2N.

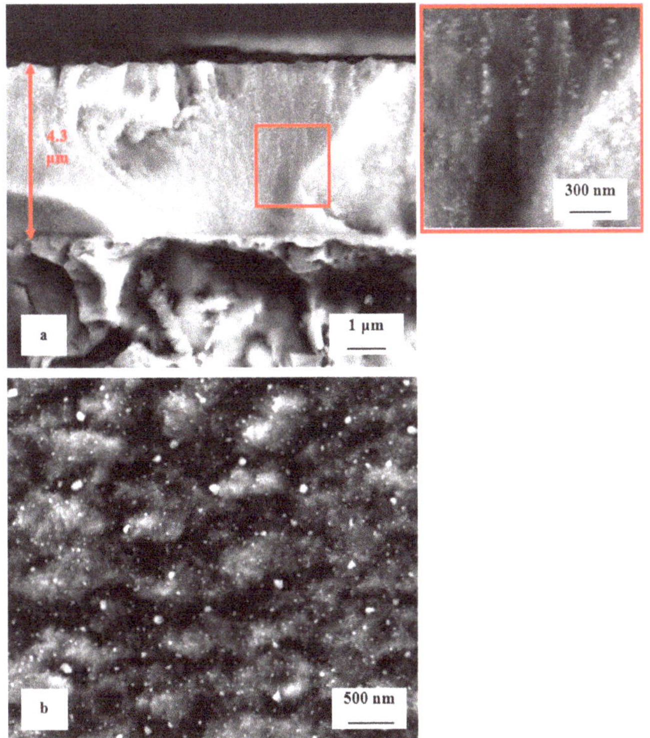

Figure 13. Microstructure of Cr_2N–6Ag film deposited on the Vanadis 6 tool steel. (a) Cross-sectional SEM micrograph, overview, detail; (b) plan-view SEM micrograph.

The mechanical properties of the Cr_2N–6Ag film are summarized in Table 4. As reported recently [39,41], the addition of silver does not influence the mechanical properties negatively in comparison with pure Cr_2N. This was attributed to the grain refinement caused by presence of Ag in the Cr_2N compound, which more than sufficiently compensated the fact that both the hardness and the Young modulus of the silver are much lower than the values normally obtained for pure Cr_2N.

Table 4. Mechanical properties of the developed Cr_2N–6Ag coating.

Coating	Nano-hardness (GPa)	Young´s Modulus (GPa)
Cr2N-6Ag	16.17 ± 1.93	263 ± 17

The flexural strength values of no-coated and Cr_2N–6Ag-coated Vanadis 6 steel manifested only marginal differences (Figure 14). The flexural strength of no-coated specimens was 4071 ± 154 MPa, while that of coated ones was 3760 ± 185 MPa. This is because there is almost no diffusion interface between the substrate and the coating, as indicated in Figure 15. Insignificant differences in flexural strength values are reflected in the values of work of fracture (Figure 16). Even though the mean value of no-coated specimens seems to be slightly higher than what was obtained for the coated steel, the statistical uncertainty ranges for both material states considerably overlap, suggesting that the work of the fracture is only slightly influenced by the presence of a PVD coating. The mentioned results are also reflected in the fracture surfaces of the Cr_2N–6Ag coated specimens. The fracture surface manifests well-visible dimple morphology, with local micro-plastic deformation at the carbide/matrix interfaces (Figure 17). This propagation manner is typical for hard materials like tool steels, as reported recently [3], and is associated with enhanced fracture surface roughness and an increase of the ductile microvoid coalescence micro-mechanism in fracture surfaces for these microstructures. One can also claim that the fracture surface of Cr_2N–6Ag coated steel does not differ significantly from the ones obtained by the flexural strength testing of no-coated steel (Figure 5).

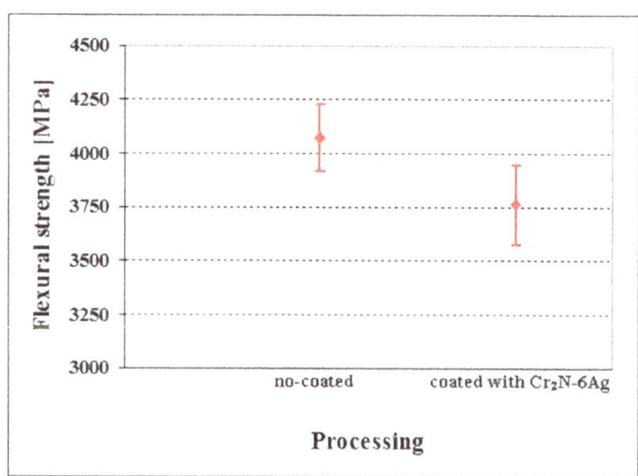

Figure 14. Flexural strength of no-coated and Cr_2N–6Ag-coated Vanadis 6 steel. The values of flexural strength are adapted from the corresponding literature [59].

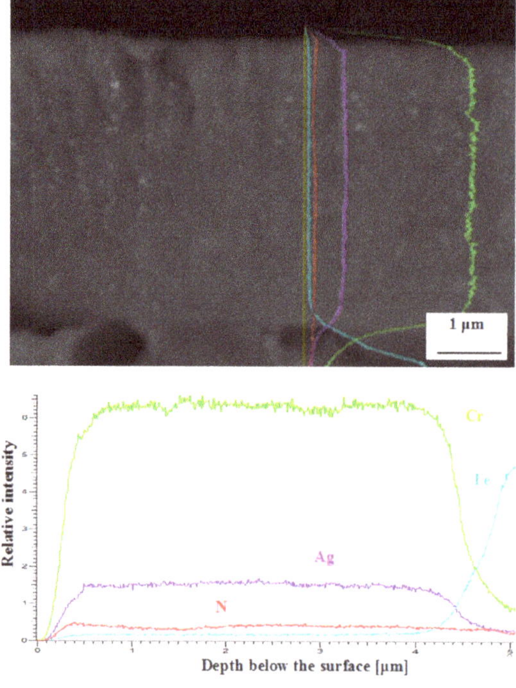

Figure 15. Concentration depth profiles of iron and the main coating elements, indicating a non-diffusive coating/substrate interface.

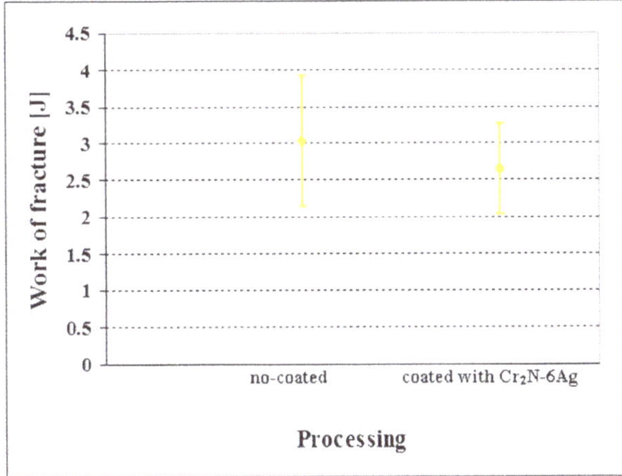

Figure 16. Work of fracture of no-coated and Cr_2N–6Ag-coated Vanadis 6 steel. The values are adapted from the corresponding literature [59].

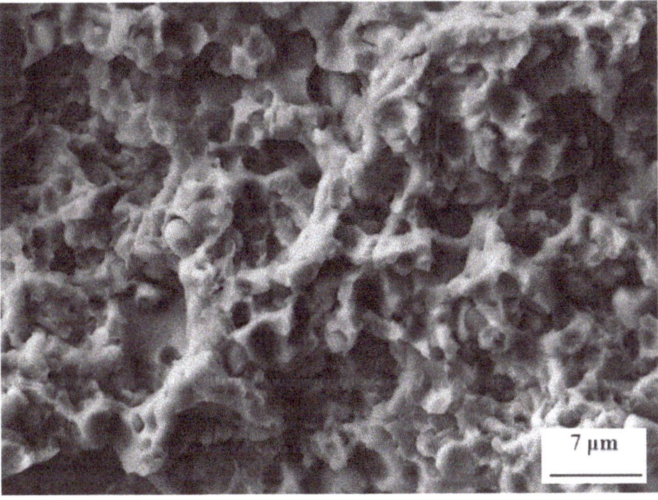

Figure 17. SEM micrographs showing the fracture surface of Cr_2N–6Ag-coated Vanadis 6 steel.

Even though a huge number of tools are coated with thin ceramic films over Europe, the information on their damages due to presence of these films on their surfaces is lacking. Furthermore, a personal experience of the author of the current paper indicates that coated tools made of ledeburitic and/or high speed steels are damaged due to poor steel quality rather than as a result of the coating presence on the surface. One can thus suggest that the coaters as well as the end-users of coated tools do not risk tool embrittlement due to PVD coating, and that the only risk for tool damage is the choice of the substrate material with poor metallurgical quality.

4. Conclusions

The obtained and summarized results on the effect of different surface engineering techniques on bulk toughness of hard and brittle Cr–V ledeburitic tool steel Vanadis 6 unambiguously indicate that:

Enhanced surface roughness has unfavorable impacts on the toughness of the material, and thereby on the durability of tools, particularly when they are heavily and dynamically loaded. It should be underlined that elevated surface roughness of tools is transmitted to the surfaces of workpieces, which leads to worsened properties of workpieces.

The presence of diffusion nitrided regions on the steel surface reduces the toughness of the material considerably. The detrimental effect of nitrided regions is evident particularly in the cases when either a surface "white" layer is formed or continuous nitrides appear at the boundaries of primary austenite grains.

PVD coating has almost no diffusion, and hence, its effect on the material toughness can be classified as marginal. Hence, the risk of material embrittlement due to the deposition of PVD thin films is minimal.

The abovementioned facts have clear consequences for the designers, heat treaters, and end-users of the tool products. While the presence of thin ceramic films does not influence the mechanical properties of the tools negatively, the enhanced surface roughness as well as the presence of nitrided regions reduces the material toughness. In nitriding, therefore, the heat treaters must inevitably avoid the formation of too thick regions, or reduce the probability of the growth of compound "white" layers on the surface. The tool designers, on the other hand, should keep in mind that the surface finish quality (roughness) influences not only the toughness of the tool itself but also the quality of the workpieces.

Author Contributions: Funding acquisition, P.J.; investigation, P.J.; supervision, P.J.; writing—original draft, review, editing, P.J.

Funding: The author acknowledges that the paper is a result of experimental works realized within the project VEGA 1/0264/17. Furthermore, this paper is a result of the project implementation "Center for Development and Application of Advanced Diagnostic Methods in Processing of Metallic and Non-Metallic Materials—APRODIMET", ITMS: 26220120014, supported by the Research & Development Operational Programme funded by the ERDF.

Conflicts of Interest: The author declares no conflict of interest.

References

1. Berns, H.; Broeckmann, C.; Weichert, D. Fracture of Hot Formed Ledeburitic Chromium Steels. *Eng. Fract. Mech.* **1997**, *58*, 311–325. [CrossRef]
2. Nemec, M.; Jurči, P.; Kosnáčová, P.; Kučerová, M. Evaluation of structural isotropy of Cr-V ledeburitic steel made by powder metallurgy of rapidly solidified particles. *Kovove Mater.* **2016**, *54*, 453–462. [CrossRef]
3. Sobotová, J.; Jurči, P.; Dlouhý, I. The effect of subzero treatment on microstructure, fracture toughness, and wear resistance of Vanadis 6 tool steel. *Mater. Sci. Eng.* **2016**, *652*, 192–204. [CrossRef]
4. Amarante, J.E.; Pereira, M.V.; de Souza, G.M.; Alves, M.F.; Simba, B.G.; dos Santos, C. Roughness and its effects on flexural strength of dental yttria-stabilized zirconia ceramics. *Mater. Sci. Eng.* **2019**, *739*, 149–157. [CrossRef]
5. Hallmann, L.; Ulmer, P.; Wille, S.; Polonskyi, O.; Köbel, S.; Trottenberg, T.; Bornholdt, S.; Haase, F.; Kersten, H.; Kern, M. Effect of surface treatments on the properties and morphological change of dental zirconia. *J. Prosthet. Dent.* **2016**, *115*, 341–349. [CrossRef]
6. Spies, H.-J.; Riese, A.; Hoffmann, W. Einflüsse auf das Bruchverhalten ledeburitischer Werkzeugstähle. *Neue Hütte* **1990**, *35*, 96–100. (In German)
7. Genel, K.; Demirkol, M.; Çapa, M. Effect of ion nitriding on fatigue behaviour of 4140 steel. *Mater. Sci. Eng.* **2000**, *279*, 207–216. [CrossRef]
8. Uma Devi, M.; Chakraborty, T.K.; Mohanty, O.N. Wear behaviour of plasma nitrided tool steel. *Surf. Coat. Technol.* **1999**, *116–119*, 212–221. [CrossRef]
9. Tier, M.; Bloyce, A.; Bell, T.; Strohaecker, T. Wear of plasma nitrided high speed steel. *Surf. Eng.* **1998**, *14*, 223–227. [CrossRef]

10. Dong, H.; Qi, P.Y.; Li, X.Y.; Llewellyn, R.J. Improving the erosion–corrosion resistance of AISI 316 austenitic stainless steel by low-temperature plasma surface alloying with N and C. *Mater. Sci. Eng.* **2006**, *431*, 137–145. [CrossRef]

11. Rousseau, A.F.; Partridge, J.G.; Mayes, E.L.H.; Toton, J.T.; Kracica, M.; McCulloch, D.G.; Doyle, E.D. Microstructural and tribological characterisation of a nitriding/TiAlN PVD coating duplex treatment applied to M2 High Speed Steel tools. *Surf. Coat. Technol.* **2015**, *272*, 403–408. [CrossRef]

12. Podgornik, B.; Hogmark, S.; Sandberg, O.; Leskovšek, V. Wear resistance and anti-sticking properties of duplex treated forming tool steel. *Wear* **2003**, *254*, 1113–1121. [CrossRef]

13. Braga, D.; Dias, J.P.; Cavaleiro, A. Duplex treatment: W–Ti–N sputtered coatings on pre-nitrided low and high alloy steels. *Surf. Coat. Technol.* **2006**, *200*, 4861–4869. [CrossRef]

14. Faga, M.G.; Settineri, L. Innovative anti-wear coatings on cutting tools for wood machining. *Surf. Coat. Technol.* **2006**, *201*, 3002–3007. [CrossRef]

15. Leskovšek, V.; Podgornik, B.; Jenko, M. A PACVD duplex coating for hot-forging applications. *Wear* **2009**, *266*, 453–460. [CrossRef]

16. Jurči, P.; Hudáková, M. Wear Mechanism of Duplex-Coated P/M Vanadis 6 Ledeburitic Steel. *Mater. Tehnol.* **2008**, *42*, 197–202.

17. Bell, T.; Dong, H.; Sun, Y. Realising of potential of duplex surface engineering. *Tribol. Int.* **1998**, *31*, 127–137. [CrossRef]

18. Pellizzari, M. High temperature wear and friction behaviour of nitrided, PVD-duplex and CVD coated tool steel against 6082 Al alloy. *Wear* **2011**, *271*, 2089–2099. [CrossRef]

19. Höck, K.; Leonhardt, G.; Bücken, B.; Spies, H.-J.; Larisch, B. Process technological aspects of the production and properties of in situ combined plasma-nitrided and PVD hard-coated high alloy tool steels. *Surf. Coat. Technol.* **1995**, *74–75*, 339–344. [CrossRef]

20. Kwietniewski, C.; Fontana, W.; Moraesa, C.; da S. Rocha, A.; Hirsch, T.; Reguly, A. Nitrided layer embrittlement due to edge effect on duplex treated AISI M2 high-speed steel. *Surf. Coat. Technol.* **2004**, *179*, 27–32. [CrossRef]

21. Vaz, F.; Rebouta, L.; Andritschky, M.; da Silva, M.F.; Soares, J.C. The effect of the addition of Al and Si on the physical and mechanical properties of titanium nitride. *J. Mater. Process. Technol.* **1999**, *92–93*, 169–176. [CrossRef]

22. Otani, Y.; Hofmann, S. High temperature oxidation behaviour of $(Ti_{1-x}Cr_x)N$ coatings. *Thin Solid Films* **1996**, *287*, 188–192. [CrossRef]

23. Munz, W.D. Titanium aluminum nitride films: A new alternative to TiN coatings. *J. Vac. Sci. Technol.* **1986**, *4*, 2717–2725. [CrossRef]

24. Hsieh, J.H.; Liang, C.; Yu, C.H.; Wu, W. Deposition and characterization of TiAlN and multi-layered TiN/TiAlN coatings using unbalanced magnetron sputtering. *Surf. Coat. Technol.* **1998**, *108–109*, 132–137. [CrossRef]

25. Gahlin, R.; Bromark, M.; Hedenquist, P.; Hogmark, S.; Hakanson, G. Properties of TiN and CrN coatings deposited at low temperature using reactive arc-evaporation. *Surf. Coat. Technol.* **1995**, *76–77*, 174–180. [CrossRef]

26. Tricoteaux, A.; Jouan, P.Y.; Guerin, J.D.; Martinez, J.; Djouadi, A. Fretting wear properties of CrN and Cr_2N coatings. *Surf. Coat. Technol.* **2003**, *174–175*, 440–443. [CrossRef]

27. Cunha, L.; Andritshky, M. Residual stress, surface defects and corrosion resistance of CrN hard coatings. *Surf. Coat Technol.* **1999**, *111*, 158–162. [CrossRef]

28. Mayrhofer, P.H.; Willmann, H.; Mitterer, C. Oxidation kinetics of sputtered Cr–N hard coatings. *Surf. Coat. Technol.* **2001**, *146–147*, 222–228. [CrossRef]

29. Lousa, A.; Romero, J.; Martinez, E.; Esteve, J.; Montala, F.; Carreras, L. Multilayered chromium/chromium nitride coatings for use in pressure die-casting. *Surf. Coat. Technol.* **2001**, *146–147*, 268–273. [CrossRef]

30. Kondo, A.; Oogami, T.; Sato, K.; Tanaka, Y. Structure and properties of cathodic arc ion plated CrN coatings for copper machining cutting tools. *Surf. Coat. Technol.* **2004**, *177–178*, 238–244. [CrossRef]

31. Nouveau, C.; Jorand, E.; Deces-Petit, C.; Labidi, C.; Djouadi, M.A. Influence of carbide substrates on tribological properties of chromium and chromium nitride coatings: Application to wood machining. *Wear* **2005**, *258*, 157–165. [CrossRef]

32. Nouveau, C.; Djouadi, M.A.; Deces-Petit, C.; Beer, P.; Lambertin, M. Influence of Cr_xN_y coatings deposited by magnetron sputtering on tool service life in wood processing. *Surf. Coat. Technol.* **2001**, *142–144*, 94–101. [CrossRef]

33. Mercs, D.; Bonasso, N.; Naamane, S.; Bordes, J.M.; Coddet, C. Mechanical and tribological properties of Cr–N and Cr–Si–N coatings reactively sputter deposited. *Surf. Coat. Technol.* **2005**, *200*, 403–407. [CrossRef]

34. Odén, M.; Almer, J.; Hakansson, G.; Olsson, M. Microstructure–property relationships in arc-evaporated Cr–N coatings. *Thin Solid Films* **2000**, *377–378*, 407–412. [CrossRef]

35. Aouadi, S.M.; Schultze, D.M.; Rohde, S.L.; Wong, K.C.; Mitchell, K.A.R. Growth and characterization of Cr_2N/CrN multilayer coatings. *Surf. Coat. Technol.* **2001**, *140*, 269–277. [CrossRef]

36. Mayrhofer, P.H.; Tischler, G.; Mitterer, C. Microstructure and mechanical/thermal properties of Cr–N coatings deposited by reactive unbalanced magnetron sputtering. *Surf. Coat. Technol.* **2001**, *142–144*, 78–84. [CrossRef]

37. Broszeit, E.; Friedrich, C.; Berg, G. Deposition, properties and applications of PVD Cr_xN coatings. *Surf. Coat. Technol.* **1999**, *115*, 9–16. [CrossRef]

38. Lamastra, F.R.; Leonardi, F.; Montanari, R.; Casadei, F.; Valente, T.; Gusmano, G. X-ray residual stress analysis on CrN/Cr/CrN multilayer PVD coatings deposited on different steel substrates. *Surf. Coat. Technol.* **2006**, *200*, 6172–6175. [CrossRef]

39. Bílek, P.; Jurči, P.; Podgornik, B.; Jenko, D.; Hudáková, M.; Kusý, M. Study of Ag transport in $Cr_2N_{0.61}$-7Ag nanocomposite thin film due to thermal exposition. *Appl. Surf. Sci.* **2015**, *357*, 317–327. [CrossRef]

40. Mulligan, C.P.; Gall, D. CrN-Ag self-lubricating hard coatings. *Surf. Coat. Technol.* **2005**, *200*, 1495–1500. [CrossRef]

41. Bílek, P.; Jurči, P.; Hudáková, M.; Pašák, M.; Kusý, M.; Bohovičová, J. Cr_2N-7Ag nanocomposite thin films deposited on Vanadis 6 tool steel. *Appl. Surf. Sci.* **2014**, *307*, 13–19. [CrossRef]

42. Jurči, P.; Bílek, P.; Podgornik, B. $Cr_2N_{0.62}$-11Ag adaptive nanocomposite thin films: Transport of Ag solid lubricant during annealing in a closed-air atmosphere. *Thin Solid Films* **2017**, *639*, 127–136. [CrossRef]

43. Carlsson, P.; Olsson, M. PVD coatings for sheet metal forming processes—A tribological evaluation. *Surf. Coat. Technol.* **2006**, *200*, 4654–4663. [CrossRef]

44. Jurči, P.; Dománková, M.; Čaplovič, L.; Ptačinová, J.; Sobotová, J.; Salabová, P.; Prikner, O.; Šuštaršič, B.; Jenko, D. Microstructure and hardness of sub-zero treated and no tempered P/M Vanadis 6 ledeburitic tool steel. *Vacuum* **2015**, *111*, 92–101. [CrossRef]

45. Jurči, P.; Dlouhý, I. Fracture Behaviour of P/M Cr-V Ledeburitic Steel with Different Surface Roughness. *Materiálové Inžinierstvo/Mater. Eng.* **2011**, *18*, 36–43.

46. Bayoumi, M.R.; Abdellatif, A.K. Effect of Surface Finish on Fatigue Strength. *Eng. Fract. Mech.* **1995**, *51*, 861–870. [CrossRef]

47. Javidi, A.; Rieger, U.; Eichlseder, W. The effect of machining on the surface integrity and fatigue life. *Int. J. Fatigue* **2008**, *30*, 2050–2055. [CrossRef]

48. Sasahara, H. The effect on fatigue life of residual stress and surface hardness resulting from different cutting conditions of 0.45%C steel. *Int. J. Mach. Tool. Manuf.* **2005**, *45*, 131–136. [CrossRef]

49. Noll, G.C.; Erickson, G.C. Allowable stresses for steel members of finite life. *Proc. Soc. Experts Stress Anal.* **1948**, *5*, 132–143.

50. Taylor, D.; Clancy, O.M. Fatigue performance of machined surfaces. *Fatigue Fract. Eng. Mater. Struct.* **1991**, *14*, 329–336. [CrossRef]

51. Novovič, D.; Dewes, R.C.; Aspinwall, D.K.; Voice, W.; Bowen, P. The effect of machined topography and integrity on fatigue life. *Int. J. Mach. Tool. Manuf.* **2004**, *44*, 125–134. [CrossRef]

52. Jurči, P.; Hudáková, M. Diffusion Boronizing of H11 Hot Work Tool Steel. *J. Mater. Eng. Perform.* **2011**, *20*, 1180–1187. [CrossRef]

53. Oliveira, C.K.N.; Casteletti, L.C.; Lombardi Neto, A.; Totten, G.E.; Heck, S.C. Production and Characterization of Boride Layers on AISI D2 Tool Steel. *Vacuum* **2010**, *84*, 792–796. [CrossRef]

54. Rodriguez-Baracaldo, R.; Benito, J.A.; Puchi-Cabrera, E.S.; Staia, M.H. High temperature wear resistance of (TiAl)N PVD coating on untreated and gas nitrided AISI H13 steel with different heat treatments. *Wear* **2007**, *262*, 380–389. [CrossRef]

55. Hnilica, F.; Čmakal, J.; Jurči, P. Changes to the Fracture Behaviour of the Cr–V Ledeburitic Steel Vanadis 6 as a Result of Plasma Nitriding. *Mater. Tehnol.* **2004**, *38*, 263–268.

56. Schell, N.; Petersen, J.H.; Bøttiger, J.; Mücklich, A.; Chevallier, J.; Andreasen, K.P.; Eichhorn, F. On the development of texture during growth of magnetron-sputtered CrN. *Thin Solid Films* **2003**, *426*, 100–110. [CrossRef]

57. Shah, H.N.; Jayaganthan, R.; Kaur, D.; Chandra, R. Influence of sputtering parameters and nitrogen on the microstructure of chromium nitride thin films deposited on steel substrate by direct-current reactive magnetron sputtering. *Thin Solid Films* **2010**, *518*, 5762–5768. [CrossRef]

58. Beger, M.; Jurči, P.; Grgač, P.; Mečiar, S.; Kusý, M. Cr_xN_y coatings prepared by magnetron sputtering method. *Kovove Mater.* **2013**, *51*, 1–10.

59. Jurči, P.; Dlouhý, I. Coating of Cr-V Ledeburitic Steel with CrN containing a small Addition of Ag. *Appl. Surf. Sci.* **2011**, *257*, 10581–10589. [CrossRef]

MDPI

St. Alban-Anlage 66

4052 Basel

Switzerland

Tel. +41 61 683 77 34

Fax +41 61 302 89 18

www.mdpi.com

Materials Editorial Office

E-mail: materials@mdpi.com

www.mdpi.com/journal/materials